Mama & Daddy courting beneath
Altamaha River bridge.
1949 Chevrolet.

Steve is two weeks old here.

Daddy, Mama, Steve, Dell
at an Apostolic Church
in Jacksonville

The man who took this picture,
E. D. McCool, was a traveling
photographer who lived in the
school bus that is behind us.

me and Kay

Daddy with the pond scoggin.
Notice splint on left leg.

Ecology of a Cracker Childhood

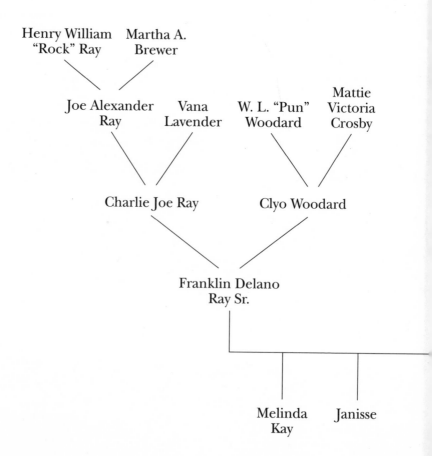

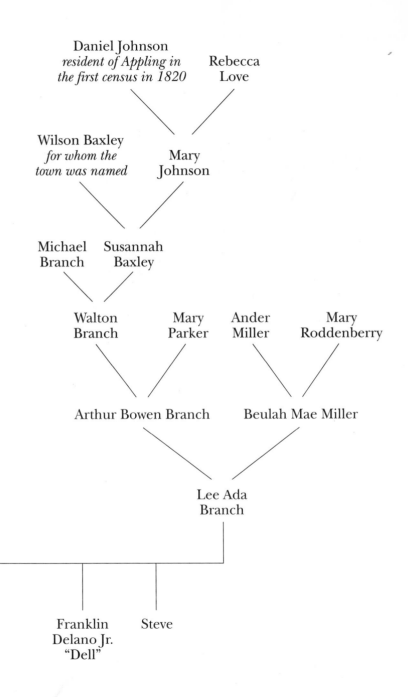

The wrecking yard in 1962. Cars
were placed in long lines. Mama is
holding me. Kay is on the blanket.

Ecology of a Cracker Childhood

JANISSE RAY

MILKWEED EDITIONS

(800) 520-6455; www.milkweed.org

Distributed by Publishers Group West

Published 1999 by Milkweed Editions
Printed in the United States of America
Jacket design by Rob Dewey
Jacket painting of the longleaf pine by David T. Williams
Jacket photographs provided by author from family album
Author photograph on jacket flap by Gary Graham
Interior design by Donna Burch
Interior illustrations by Bruce Lyndon Cunningham
Endsheet photographs and photograph on p. iv provided by author from family album
The text of this book is set in New Baskerville
    00 01 02 03 5 4 3
First Edition

The book epigraph on p. x is from Iain Crichton Smith, "Shall Gaelic Die?" in *The Faber Book of Twentieth-Century Scottish Poetry,* ed. Douglas Dunn (London: Faber and Faber, 1992).

The epigraph on p. 267 is from James Dickey, "Cherrylog Road," in *The Whole Motion: Collected Poems* (Middletown, Conn.: Wesleyan University Press, 1992). Copyright © 1992 by James Dickey. Reprinted with permission from University Press of New England.

The appendixes on pp. 275–84 are adapted from Reed F. Noss, Edward T. LaRoe III, and J. Michael Scott, *Endangered Ecosystems of the United States: A Preliminary Assessment of Loss and Degradation* (National Biological Survey, February 1995).

Milkweed Editions, a nonprofit publisher, gratefully acknowledges support from our World As Home funders: Lila Wallace-Reader's Digest Fund; Creation and Presentation Programs of the National Endowment for the Arts; and Reader's Legacy underwriter, Elly Sturgis. Other support has been provided by the Elmer L. and Eleanor J. Andersen Foundation; James Ford Bell Foundation; Bush Foundation; Dayton Hudson Foundation on behalf of Dayton's, Mervyn's California, and Target Stores; Doherty, Rumble & Butler Foundation; General Mills Foundation; Honeywell Foundation; McKnight Foundation; Minnesota State Arts Board through an appropriation by the Minnesota State Legislature; Norwest Foundation on behalf of Norwest Bank Minnesota; Lawrence and Elizabeth Ann O'Shaughnessy Charitable Income Trust in honor of Lawrence M. O'Shaughnessy; Oswald Family Foundation; Ritz Foundation on behalf of Mr. and Mrs. E. J. Phelps Jr.; John and Beverly Rollwagen Fund of the Minneapolis Foundation; St. Paul Companies, Inc.; Star Tribune Foundation; U.S. Bancorp Piper Jaffray Foundation on behalf of U.S. Bancorp Piper Jaffray; and generous individuals.

Library of Congress Cataloging-in-Publication Data

Ray, Janisse, 1962–
        Ecology of a Cracker childhood / Janisse Ray. — 1st ed.
            p.   cm.
        ISBN 1-57131-234-X (hardcover)
            1. Ray, Janisse, 1962–   —Childhood and youth. 2. Baxley (Ga.)—
        Biography. 3. Longleaf pine. 4. Deforestation—Georgia. 5. Forest ecology—
        Georgia. 6. Consumption (Economics)—Social aspects. I. Title.
        F294.B39R39 1999
        975.8'784—dc21
        [B]   99-13403
                                                                                          CIP

This book is printed on acid-free, recycled paper.

For my son
Silas

and for the land

Ecology of a Cracker Childhood

Introduction	3
Child of Pine	5
Below the Fall Line	13
Shame	17
Built by Fire	35
Iron Man	39
Forest Beloved	65
Junkyard	71
Crackers	81
Native Genius	89
Timber	99
Heaven on Earth	105
Clearcut	123
How the Heart Opens	127
Longleaf Clan	141
Clyo	143
Hallowed Ground	151
Poverty	157
The Keystone	167
Beulahland	175
Indigo Snake	187
Mama	193

Bachman's Sparrow   205

Light   211

Flatwoods Salamander   217

Altamaha River   223

Pine Savanna   239

Driving and Singing   245

The Kindest Cut   251

Leaving   255

Second Coming   267

Afterword: Promised Land   271

*There Is a Miracle for You
    If You Keep Holding On*   273

Appendixes   275

Acknowledgments   285

Words rise out of the country.

—Iain Crichton Smith

Ecology of a Cracker Childhood

Introduction

In south Georgia everything is flat and wide. Not empty. My people live among the mobile homes, junked cars, pine plantations, clearcuts, and fields. They live among the lost forests.

The creation ends in south Georgia, at the very edge of the sweet earth. Only the sky, widest of the wide, goes on, flatness against flatness. The sky appears so close that, with a long-enough extension ladder, you think you could touch it, and sometimes you do, when clouds descend in the night to set a fine pelt of dew on the grasses, leaving behind white trails of fog and mist.

At night the stars are thick and bright as a pint jar of fireflies, the moon at full a pearly orb, sailing through them like an egret. By day the sun, close in a paper sky, laps moisture from the land, then gives it back, always an exchange. Even in drought, when each dawn a parched sun cracks against the horizon's griddle, the air is thick with water.

It is a land of few surprises. It is a land of routine, of cycle, and of constancy. Many a summer afternoon a black cloud builds to the southwest, approaching until you hear thunder and spot lightning, and even then there's time to clear away tools and bring in the laundry before the first raindrops spatter down. Everything that comes you see coming.

That's because the land is so wide, so much of it open. It's wide open, flat as a book, vulnerable as a child. It's easy to take advantage of, and yet it is also a land of dignity. It has been the way it is for thousands of years, and it is not wont to change.

I was born from people who were born from people who were born from people who were born here. The Crackers crossed the wide Altamaha into what had been Creek territory and settled the vast, fire-loving uplands of the coastal plains of southeast Georgia, surrounded by a singing forest of tall and widely spaced pines whose history they did not know, whose stories were untold. The memory of what they entered is scrawled on my bones, so that I carry the landscape inside like an ache. The story of who I am cannot be severed from the story of the flatwoods.

To find myself among what has been and what remains, I go where my grandmother's name is inscribed on a clay hill beside my grandfather. The cemetery rests in a sparse stand of remnant longleaf pine, where clumps of wiregrass can still be found. From the grave I can see a hardwood drain, hung with Spanish moss, and beyond to a cypress swamp, and almost to the river, but beyond that, there is only sky.

Child of Pine

When my parents had been married five years and my sister was four, they went out searching among the pinewoods through which the junkyard had begun to spread. It was early February of 1962, and the ewes in the small herd of sheep that kept the grass cropped around the junked cars were dropping lambs.

On this day, Candlemas, with winter half undone, a tormented wind bore down from the north and brought with it a bitter wet cold that cut through my parents' sweaters and coats and sliced through thin socks, stinging their skin and penetrating to the bone. Tonight the pipes would freeze if the faucets weren't left dripping, and if the fig tree wasn't covered with quilts, it would be knocked back to the ground.

It was dark by six, for the days lengthened only by minutes, and my father had gone early to shut up the sheep. Nights he penned them in one end of his shop, a wide, tin-roofed building that smelled both acrid and sweet, a mixture of dry dung, gasoline, hay, and grease. That night when he counted them, one of the ewes was missing. He had bought the sheep to keep weeds and snakes down in the junkyard, so people could get to parts they needed; now he knew all the animals by name and knew also their personalities. Maude was close to her time.

In the hour they had been walking, the temperature had fallen steadily. It would soon be dark. Out of the grayness Mama heard a bleating cry.

"Listen," she said, touching Daddy's big arm and stopping so suddenly that shoulder-length curls of dark hair swung across her heart-shaped face. Her eyes were a deep, rich brown, and she cut a fine figure, slim and strong, easy in her body. Her husband was over six feet tall, handsome, his forehead wide and smart, his hair thick and wiry as horsetail.

Again came the cry. It sounded more human than sheep, coming from a clump of palmettos beneath a pine. The sharp-needled fronds of the palmettos stood out emerald against the gray of winter, and the pine needles, so richly brown when first dropped, had faded to dull sienna. Daddy slid his hands—big, rough hands—past the bayonet-tipped palmetto fronds, their fans rattling urgently with his movements, him careful not to rake against saw-blade stems. The weird crying had not stopped. He peered in.

It was a baby. Pine needles cradled a long-limbed newborn child with a duff of dark hair, its face red and puckered. And that was me, his second-born. I came into their lives easy as finding a dark-faced merino with legs yet too wobbly to stand.

My sister had been found in a big cabbage in the garden; a year after me, my brother was discovered under the grapevine, and a year after that, my little brother appeared beside a huckleberry bush. From as early as I could question, I was told this creation story. If they'd said they'd found me in the trunk of a '52 Ford, it would have been more believable. I was raised on a junkyard on the outskirts of a town called Baxley, the county seat of Appling, in rural south Georgia.

In the 1970 census, Baxley listed 3,500 people and Appling County figured almost 13,000, and a decade later the figures had risen only slightly. Even in 1990 the county's population was not even 16,000, the town's a little over 3,800—hardly more than it had been thirty years earlier when I was a girl. Projections for the year 2000 don't show much of a change. The nearest bigger town was Waycross forty miles south, population 19,000, located on the northern edge of the unvanquished Okefenokee Swamp, and anything that could be called a city lay two or three hours, a hemisphere, away: Jacksonville to the south; Macon to the north; Savannah to the east, on the Atlantic. But those places were outside my knowing, and foreign.

What I knew was a 20' x 26', white, clapboard house that sat in the middle of ten brushy acres my father had newly purchased. He'd built the house and strung a hog-wire fence around it, inside which my mother had planted sapling plums and pears and outside which junk was stacked and piled. The house had two small bedrooms opening onto a short hall that joined the living room/ kitchen. A thick white sheepskin lay in front of the gas space heater in the hallway, and a little organ that no-body could play sat in the living room. Baby pictures lined the organ. The house's back steps were concrete blocks that wobbled and grumbled underneath your feet and a screened porch hung off the front door.

Mama tried to keep flowers, beds of four-o'clocks, old

maids, and daylilies around the house's drip line, and mowed the grass regularly. She arranged chairs in the yard among pretty ornaments—a ceramic frog, a tiny wrought-iron tea table. She spray-painted a cast-iron cookstove black and set it up on bricks and hung a wash-pot from a metal frame, and when the sheep got in and ate the flowers, she drove them from the yard with her broom, yelling "shoo, shoo" and "git."

As soon as I learned to walk, I would wander into the beyond, where the junk began, touching the pint-sized redbud my mother had planted or the hymn of china-berry that dropped wrinkled and poisonous drupes into the grass. When my mother called, I would crouch in the dirt behind the water pump, perfectly still and quiet, making her search for me. It wasn't a game.

"Half wild," she'd murmur. She had to tie bells on my shoes, silver jingle bells that gave away my whereabouts and led her to me.

My hair was long and stringy, snarled at the nape like a rat's nest, so that Mama, who had brushed it out just that morning, chased me with a hairbrush and pinned it out of my eyes with barrettes. My mouth was dirty as was my dress, and my feet were perpetually dirty, and Mama would pop me in the bathtub and button me into a ruffled dress dotted with flowers she had handsewn on her Singer. But she'd need to peel potatoes for supper or slice a mess of okra, and I would stand looking out the screen door until I worried it into opening, and when she found me, I would again be dirty.

When I was bigger, I could get up into the trees, es-pecially the chinaberry, which had grown quickly and notched after a couple of feet. I would sit in the tree and

wait, listening for something—a sound, a resonance—
that came from far away, from the past and from the
ground. When it came, the sun would hold its breath, the
tree would shiver, and I would leap toward the sky, hoping finally for wings, for feathers to tear loose from my
shoulders and catch against sweeps of air.

The ground was hard, unyielding, but it wanted me,
reaching out its hard, black arms and rising in welcome.
I would lift and run along it as fast as I could and think
again of soaring, of flying, until I was breathless and oily
with sweat, and then I would collapse to the earth.

I could unhook the chain from the nail on the post
and leave the yard, but I wasn't allowed to go far. The
junkyard was dangerous, strewn with broken glass and
shards of rusty metal. A rusty nail could send a streak of in-
flammation into the bloodstream, Daddy warned, causing
lockjaw that might clamp your mouth shut like a beaver
trap, even in the middle of supper, with a chicken bone
hanging out. There was no way to undo lockjaw.

All kinds of accidents could happen in the junkyard. A
bad cut would mean stitches with a thick, curved needle.
Or you could get poisoned the way Steve, my younger
brother, had when he drank motor oil. He was four or
five, hanging out with Mama and Daddy at the shop while
they worked on a motor, the rest of us at school. Daddy
had drained motor oil into an oily green Coke bottle,
and left it on the oil-stained concrete floor of the shop.
Since it looked like Coke, Steve reckoned it was, and so
sure was he, he never tasted the virulence until it was too
late. He had to have his stomach pumped out, which
meant a big hose down your throat.

Besides, the junkyard was giant enough that little girls

could get lost and not find their way out. Wild mongooses or orangutans with big yellow teeth would chase you and even catch you and who knows then what would happen. Bad men might have climbed the far fence to steal car parts and hit you over the head with an adjustable wrench and kidnap you.

All of it had to be true. My sister and I slept in the back bedroom, and one night I woke to hear the cement blocks of the back steps complaining as someone climbed them. I heard the latch of the screen door rattle. It held. The steps rasped again. Because I was too frightened to call out, I lay petrified, waiting for noises at my own casement, listening to the seconds tick by on the windup clock in the living room, hearing nothing more and still nothing except my father's snores in the next room, until finally sleep drowned me.

I dreamed I was sitting on the cold toilet, trying to pee. The bathroom was made of windows, glass on all sides, and beyond the windows was darkness. Then I saw something—a monster, a man—coming for me, out of the darkness, and I peed desperately so I could rush back into the warm quilt-swaddled bed I shared with my sister before the evil arrived. When I woke it was morning, beautiful eastern sunlight casting the room to sunflowers and the bed flooded with urine.

Almost every night I wet the bed.

One night Dell, the middle brother, woke and looked out the window that was level with his bed. He said he saw two men carrying off radiators, one in each hand. They were climbing the fence, leaving, when he wakened and saw them. He never thought to shake anybody or call out, although the next morning he crossed his heart and hoped to die if what he said wasn't true. Radiators were stacked here and there, scrambled in with all the other junk, so there was no way of counting those left to see if any were missing, nor could we find footprints. No matter, I believed him. Hadn't I, not long before, heard the screen door rattle in the middle of the night?

Daddy bought a piano even though we had an organ, stuffed it in our undersized house and started sending my sister and me to piano lessons. Mrs. Mobley taught in a cold concrete building near the elementary school, and we walked there after school one day a week for a thirty-minute lesson. Kay did well, but struggle as I might, I could not learn quarter notes and "Every Good Boy Does Fine" and four-four time. I hated piano.

"Please let me quit," I begged my father.

"Playing an instrument is something you'll appreciate your whole life," he said. "I want you to stick with it." He made me practice thirty minutes a day.

The next week I would beg again. "Please. I hate piano. . . . I don't understand any of it." He would not let me quit.

Finally I said contentiously, trembling at my rare boldness, "You're wasting money on me. I can't learn to play piano. I want to be outside." Maybe my teacher whispered to him that I had no aptitude or perhaps having one musical daughter was enough, given the scarcity of funds—many weeks he could barely find the money to pay Mrs. Mobley—but that week's lesson was my last.

I would rather be sitting in this certain pine tree I loved. It was within hollering distance of the house and eyeshot of the shop, and I was allowed to go to it when we

weren't working, before homework, and even if I hadn't
been allowed, I would've sneaked there anyway. It was a
good pine, forty-feet tall, sturdy and easy to climb if you
boosted up from the fender of a '59 Chevy beneath it that
sat on long-flat tires, collecting straw against its wiper
blades, in the trenches of its bumpers and along its flat
trunk. Blue jays criticized each other in the tree and fussed
at me as I rode the lowest pine limb as if on horseback, not
knowing enough about anything but eager to live, listening to the wind in the needles that was sufficient music.

Almost every summer afternoon a thunderstorm
would build in a corner of the sky until it burst and, in
its bursting, alleviate a portion of the intense heat and
humidity of southern Georgia, lightening the barometric
pressure so people forgot to be irritable. Thunder would
clap and rattle the sky, followed by strikes of lightning
that tore clouds open like paper bags. Rain gushed down.
Just before a storm hit, the wind would pick up—it's not
at all a windy place, hardly even a breeze usually—and
the cooling winds would be satisfyingly strong.

I would run for a tree, even the thirty-foot chinaberry
in the yard, and climb halfway, straddling a stout limb
and facing the wind that whipped my face and billowed
out my skirt, pushing against me. The tree shook and
swayed and I hung on, laughing for what I knew as joy.
There is something about storms. Maybe it's that wind
begs for resistance. If Mama saw me, she would yell from
the doorway to get down out of that tree and lecture later
that it wasn't ladylike, me climbing trees with a skirt on,
skinning up my legs, and that I would tear my clothes or,
even worse, fall, if the limb didn't break first or if lightning didn't strike me.

Below the Fall Line

The landscape that I was born to, that owns my body: the uplands and lowlands of southern Georgia. The region lies below what's called the fall line, a half-imaginary demarcation avouched by a slight dip in the land, above which the piedmont climbs to the foothills of the Blue Ridge, then up that mountain chain to the eastern continental divide. The fall line separates the piedmont from the Atlantic coastal plain—a wide flat plateau of pineywoods that sweeps to a marble sea.

My homeland is about as ugly as a place gets. There's nothing in south Georgia, people will tell you, except straight, lonely roads, one-horse towns, sprawling farms, and tracts of planted pines. It's flat, monotonous, used-up, hotter than hell in summer and cold enough in winter that orange trees won't grow. No mountains, no canyons, no rocky streams, no waterfalls. The rivers are muddy, wide and flat, like somebody's feet. The coastal plain lacks the stark grace of the desert or the umber panache of the pampas. Unless you look close, there's little majesty.

It wasn't always this way. Even now in places, in the Red Hills near Thomasville, for example, and on Fort

Stewart Military Reservation near Hinesville, you can see how south Georgia used to be, before all the old longleaf pine forests that were our sublimity and our majesty were cut. Nothing is more beautiful, nothing more mysterious, nothing more breathtaking, nothing more surreal.

Longleaf pine is the tree that grows in the upland flat-woods of the coastal plains. Miles and miles of longleaf and wiregrass, the ground cover that coevolved with the pine, once covered the left hip of North America— from Virginia to the Florida peninsula, west past the Mississippi River: longleaf as far in any direction as you could see. In a longleaf forest, miles of trees forever fade into a brilliant salmon sunset and reappear the next dawn as a battalion marching out of fog. The tip of each needle carries a single drop of silver. The trees are so well spaced that their limbs seldom touch and sunlight streams between and within them. Below their flattened branches, grasses arch their tall, richly dun heads of seeds, and orchids and lilies paint the ground orange and scarlet. Purple liatris gestures across the landscape. Our eyes seek the flowers like they seek the flashes of birds and the careful crossings of forest animals.

You can still see this in places.

Forest historians estimate that longleaf covered 85 of the 156 million acres in its southeastern range. By 1930, virtually all of the virgin longleaf pine had been felled. Now, at the end of the twentieth century, about two mil-lion acres of longleaf remain. Most is first- and second-growth, hard-hit by logging, turpentining, grazing, and the suppression of fire.

Less than 10,000 acres are virgin—not even 0.001 per-cent of what was. There's none known in Virginia, none in Louisiana, none in Texas, none in South Carolina. About 200 old-growth acres remain in Mississippi, about 300 in Alabama, and almost 500 in North Carolina, in four separate tracts. The rest survives in Georgia and Florida. An estimated 3,000 acres of old-growth in Georgia lie on private land, precariously, and the largest holding of virgin longleaf, about 5,000 acres, belongs to Eglin Air Force Base in Florida.

In a 1995 National Biological Service assessment of biological loss, ecologist Reed Noss classified the longleaf/ wiregrass community as "critically endangered." Ninety-eight percent of the presettlement longleaf pine barrens in the southeastern coastal plains were lost by 1986, he said. Natural stands—meaning not planted—have been reduced by about 99 percent.

Apocalyptic.

This was not a loss I knew as a child. *Longleaf* was a word I never heard. But it is a loss that as an adult shadows every step I take. I am daily aghast at how much we have taken, since it does not belong to us, and how much as a people we have suffered in consequence.

Not long ago I dreamed of actually cradling a place, as if something so amorphous and vague as a region, existing mostly in imagination and idea, suddenly took form. I held its shrunken relief in my arms, a baby smelted from a plastic topography map, and when I gazed down into its face, as my father had gazed into mine, I saw the pine flatwoods of my homeland.

Shame

I dont hate it. . . . I dont hate it. . . . I dont hate it *he thought, panting in the cold air, the iron New England dark;* I dont. I dont! I dont hate it! I dont hate it!

—William Faulkner, *Absalom, Absalom*

A junkyard wasn't a bad place to grow up. It was weird enough to stoke any child's curiosity, a playground of endless possibility. Since my two brothers and I were doorstep kids, which is what you call siblings born one after the other, little more than a gestation period apart, the three of us made a pack. We taught school to each other alongside a '50 Nash the bluish green color of a chalkboard, using stubs of chalk filched from school to write sums and spelling words on the car body, wiping them away with a rag eraser. Letters got scribbled across the sides of the Nash, around door handles, skipping the crack where the door closed into the car body, ending at windshields. When my brothers would not have the patience to pretend school after enduring it all week, I lined my dolls on the ground and taught them.

"*Rides* is the verb, and *girl* is the noun." I would be standing by the Nash, hands powdered white. This day my brothers were playing. Dell had found an actual school desk left out in the weather and not in very good shape, wood laminations peeling away initials carved into it, and he was sitting at it; Steve was atop a wooden wire spool he had rolled across the wet dirt.

"Jesse, what part of speech would *horse* be?" I asked.

There was no answer, of course.

"He's absent," Steve said.

"Okay, Stephen," I said. "Then you tell me."

"That would be a pile of doo-doo," he said. We laughed.

"Young man, you are not going to use such language in my classroom," I said sternly. "Immediately give me the correct answer or you're in deep trouble."

"That would be deep doo-doo," he said.

"Teacher, that sentence looks like a pile of crap to me," Dell chimed in.

"Your mouth is despicable," I said. "You are not allowed to use filthy language in my classroom. I am of good mind to wash out your mouths with lye soap." We imitated our worst teachers, saying things we dared not say in real school.

"We'll say what we dern well please," Steve said.

"I'm going to have to reprimand both of you." I was talking through my nose. "Both of you terrible children come to the front of the room this minute."

"What room?" Dell said, and we all laughed. They wouldn't come voluntarily, so I went back and grabbed them by their shirts and dragged them across the sparse grass—sheep-crop close to the ground. I told them I was going to spank their hands with my ruler.

The ruler, unfortunately, was in the backseat of the car, a dusty, cobwebby, moldy, flaky closet, and I could not hold the wriggling pupils and retrieve the ruler at the same time. Before I could reach the ruler, Dell had run, jumping the ditch and heading off into the junkyard, yelling taunts.

"Poo-poo teacher! Poo-poo school!"

"You come back here this minute," I yelled. Now Steve took off running, and I chased him, brandishing the wooden ruler, around and over junk cars, up the trunks, over the tops, down the hoods, metal popping back into shape behind us. Dell paused on one roof to stick his thumbs in his ears and waggle his fingers, tongue out. This was one of our favorite games. Mama could hear us all the way to the house. I remember standing on many a hood, judging the distance between me and the next car and swallowing at the gap of space but knowing I could make it if I dared. That was good practice for my life.

If we missed and sliced our knees and feet, Mama would tear bandages out of clean white sheets, doctor us with an old-time tree salve called Balsam Peru, and tape us up. If the cut was on the foot, she might make a sandal out of a cloth square the length of our foot, split it as if making a wide fringe on either side of the foot, and tie the bandage on. We looked like Renaissance younguns with these white cloth sandals, and we would ask for bandages on both feet even if only one was cut. Then the others would want sandals too.

When we got tired of chasing, we decided to play preacher awhile, climbing the sheet-metal steps of the breached school bus and finding seats in the nave. Today Dell would preach and Steve would be baptized.

None of us had been baptized yet, but we'd seen others climbing into the wooden tank at the side of the church, wearing long white robes, and Brother Randolph praying over them in the waist-deep water. Then they would go under and come up streaming wet, and the sisters would fling white towels around them, for the robes would be clinging to their bodies, and they would leave a trail of water, like a slug, across the floor of the church as they hastened to change into dry clothes. They wouldn't look any different when they came up than before going under, but I supposed they'd changed inside. Water worked an invisible magic.

Our baptism pool was at the back of the bus, and we had to pretend it was full of water. We sang "Shall We Gather at the River." Nobody knew any of the words except the chorus, which we sang over and over: "the beauti-ful, beau-ti-ful riv-er." Dell prayed, and it came out as a joke, even if we didn't mean it to. "Forgive this man his sins, and let him do good work in the world." Then he knocked Steve back on the hard, dusty bench. Steve spluttered and came up coughing.

"Preacher, you nearly drowned me."

"Brother, we were washing your sins away," Dell replied.

"But I think I have brain damage."

Late afternoon I'd make my way back to the Nash to put away the chalk and erasers and wash the dusty sides with water from the ditch. The door was still standing open the way I had left it, and when I closed it, my sentence, which had been written across the crack, was made whole again. One of my dolls might behave well enough to be teacher's helper in washing the board.

Or I would leave the board filled with writing, knowing that a thunderstorm loomed and rain would do the work.

The junkyard was stuffed with junked, wrecked, rusted, burned, and outmoded automobiles and parts of automobiles. Ragweed and dog fennel sort of hid the mess from the road, but when you turned off the highway, you got the full effect. It was like sticking your head into a wide-angle trash can.

Mercury, Ford station wagon, Cadillac, Rambler, Nova, Mustang, Dodge Dart, Chevy truck, Chevy van, '55 Crown Victoria. Volkswagen, '55 Chevrolet, Nash, Hudson. Bread truck, military milling shop on wheels, school bus, Farmall tractor. John Deere, moving van. Mazda pickup, Gremlin, Falcon, Metropolitan.

'Thirty-nine Packard hearse, '49 Buick, BMW, Oldsmobile, President Studebaker. Galaxy 500, Fairlane 500, '26 Diamond T, Uniform Van, Cougar. Most of the cars were '30s to '60s models: '40 Ford, '39 Mercury, '47 Cadillac, Imperial, '58 Hudson Hornet with a Packard engine. Mercedez Benz.

Some of them had stories. When Mama was pregnant with Kay, they'd wrecked in the '47 Cadillac. She was thrown through the windshield and needed stitches on her gushing head and on her right elbow, where she had landed. When the baby Kay was found, she bore a birthmark on her right elbow.

The Golden Hawk Studebaker had come from Foster Sellers, bank robber. From Baxley, he was on the FBI's "Top Ten Most Wanted" list. He was notorious for robbing banks and not getting caught and for escaping prison when he did. Once he was picked up in Baxley, reported after having breezed home to visit his mother.

"Foster," the sheriff said to him on the ride to the jail. "You could be anything you want to be."

"I am what I want to be," Foster replied. Another thing he said was that when he cased out a bank, he looked up and down the street for a police officer. If he didn't see one, he assumed there wouldn't be one.

Once, Foster's car was found stashed in the woods, covered with brush. Lawmen carrying machine guns poured into town—FBI, GBI, the Armed Guard—and immediately blocked every road leading out. They were searching even the trunks of cars. Somehow Foster gave them the slip. Late that evening, he called the sheriff's office from a long-distance phone and asked, "Have y'all found Foster yet?"

We could climb behind the wheel of the gangster Golden Hawk and be Foster Sellers, zooming away with the cash, and be whoever else we wanted.

Between car bodies were swales of scrap: bathtubs, motors, an airplane wing, the bucket to a crane, tractor tires, sparrow nests, froze-up treadle sewing machines, kids' swimming pools going to crumbs, rusted harrows and plows. More motors. Transmissions. A small mountain range of bald tires and rims. Hubcaps. A gaggle of bent, broken, mutilated, unusable bicycle frames.

Old vacuum cleaners, crates of canned goods with lids rusted so tight you'd have to break the jars to get them off, a dead refrigerator. Daddy collected hills of aluminum cans out of dumpsters and off roadsides because he intended to roof the house with aluminum shingles one day and went so far as to invent a machine that would slice off the top and bottom and split them down the center to make a curling square. Axles, transmissions, a wheelbarrow with a flat tire, split plastic buckets, half a toilet, a postmodern clay statue of a twisted man with a cigarette in his ear, wire cages, broken wood stoves, an upside-down washing machine.

Ten acres of failed machines.

In summer we would skim along the flanks of junk-piles, wearing shoes for once, gathering fat blackberries that grew healthy in the mineral-rich loam, their canes mulched with scrap iron. The blackberries were luxuriant. We couldn't get our fill.

Every couple of years a car crusher came from a Savannah foundry, either with a crane that dropped a two thousand-pound metal ball over and over, spewing glass missiles, or with a guillotine-like box that lowered a metal slab onto cars crammed inside, the way you'd press peanut butter cookies down with the tines of a fork.

A front-end loader first yanked out gas tanks that otherwise would explode in the smelter. The tanks are square and flat, and the boys and I would claim one and drag it down to the pond. It floated on its own but not well, the deck riding just below the water's surface with somebody aboard, so we'd lash on empty milk jugs and raft up and down the shallow pond, racing. We'd tie flags on the conduit arcing like a stallion's neck before us, where you'd hose gas into the tank normally, and sometimes we'd rope them all together into a Huck Finn raft and pole up and down, jumping off, climbing on. We tried to float hoods too, ones from old-model cars that curve upward. We'd tie inner tubes underneath and sit on hubcap seats. Anything to float. We couldn't wear bathing suits, and our soaked clothes would be heavy and unwieldy, clinging to our skin. By supper time we'd be coated with rust and grease, wet and dirty, and we'd have to sneak in the backdoor to escape Mama.

At one end of the pond a red maple grew, although I didn't know its name then. I only knew it was as red-veined as any human and that its leaves flung sheer gold in fall. It was considered my tree. If I didn't want anyone to climb it, he couldn't. I built a fort, a matter of a few boards and ropes, and climbed the maple to read, to think, and to watch my brothers, one and two years younger, floating the pond that has long since given way to vegetative succession and is no longer a pond. The last time I saw the tree I could barely reach the lowest limb; I couldn't imagine having climbed it.

Odd how things are and then they aren't.

I was in sixth or seventh grade when we formed the Thingfinders Club, the objective of which was to ransack cars for whatever valuables we could find under seats, in arm rests, beneath floor mats: coins, combs, jewelry. We were like burglars, wrenching open car doors from two sides and running our hands between and beneath the seats, lifting floor mats, plundering the glove compartment. A quarter was an excellent haul. Buffalo nickels and wheat pennies, coppers minted before 1958, with a wreath of wheat on the obverse side, excited us too, because we could trade them to Daddy for more than face value. We'd marvel that anyone would sell an old vehicle without cleaning it out, and we'd hold meetings down by the sheep pen to display our loot for swapping.

Every Saturday afternoon we went to town. There is no other way to say it. We went to town. Baxley was lively with families on Saturday afternoons, and it was our highlight of the week, social and otherwise, for the likelihood of not seeing friends or kinfolk was nil, although the extent of our socializing was a brief *howdy, how you doing?* and a few-minutes' chat or a shy and quickly exchanged wave with a schoolmate. There was no playing. We bathed, scrubbed, and donned clean clothes in preparation and anticipation. We washed our hair.

Mostly the purpose of the trip was groceries. Three grocery stores served the town, but Daddy shopped at Winn-Dixie because he thought it had better prices and because Daddy could offer the manager five dollars for an entire buggy-load of dented cans or day-old bread and the manager would take it. We might come home with anything: beets, cat food, mandarin oranges. Daddy was the greatest spendthrift imaginable. If he could get a truckload of bread, he'd squash it flat and put it in people's freezers to gradually feed to his thirty sheep.

Surprise is addicting, I think, and I can't tell you how fun it was to come home and unpack boxes of miscellany, a trunkful, not knowing what you might find. String beans, shoe polish, condensed milk, crushed pineapple, olives. Sometimes it was merchandise we couldn't use or didn't like (a bottle of hair dye, capers) and these items got packed away in an old car body or given away. If the labels had been torn from the cans and they were not distinctive in size, like evaporated milk or tomato paste, they occupied a special stupefying corner of the pantry to be opened when Mama was willing to serve whatever she encountered.

Other than those grab-buggy surprises, Mama

bought the same things week after week, staples: white rice, fryers, sugar, flour, cornmeal, grits, eggs, tea, a five-pound sack of potatoes, black-eyed peas, bacon, peanut butter, saltines, white bread, hamburger meat. If ice milk was on sale we would get a half gallon of the fudge ripple, which got divided into halves, then cut into six sections with a knife to make equal servings. We never ever lacked food, but we had few treats and what we had was divided meticulously. Whoever did the dividing picked his piece last.

Winn-Dixie supermarket was in a strip mall, the first and only in town, that didn't seem like one because it was downtown, two blocks from the courthouse and one street off the railroad track. In the same center was a dollar store and T,G & Y, a department store chain whose initials, we joked, stood for turtles, girdles, and yo-yos.

Every Saturday, Daddy split a dollar four ways to give us spending money. In the 1960s a quarter would buy a sackful of penny candy or a box of Cracker Jacks, or maybe we'd had a good week in the cars and could buy a set of jacks or a box of crayons. If a tractor had sold and Daddy was flush, he might hand us two dollars to divide.

Some weeks nothing sold on the junkyard, and we skipped going to town.

My brothers were still in elementary by the time they learned enough about mechanicking to get a go-cart running and managed to claim about as much time riding as they did fixing. A go-cart was like an old-model Harley; you had to work for any pleasure it offered. The boys learned to file points, change spark plugs, clean out gas filters, and pour gas in the carburetors to fire them; they learned to jerk the pull rope and rewind it when it unknotted. If they wanted to ride they had no alternative, and children want to ride.

We collected Coke bottles for gas money. Each one was worth a dime when turned in for reuse at the bottling plant across town. Ten of them bought a gallon and a half of gas. We walked the side road—if it had a name nobody knew it—ferreting bottles from ditches, emptying out the mossy, bug-filled water inside and lugging them home in a croker sack, a burlap bag that smelled of cracked corn, which Daddy sometimes bought to feed the sheep. Any bottles found in junk cars we were allowed to keep.

Each bottle had the name of the town where the soda was bottled imprinted on its base. Fitzgerald, Georgia. Savannah. Waycross. We'd finger the names of those far-away places we knew nothing of, marveling that a name could tell so little. Any we found from Baxley, which had had a bottling plant, we couldn't bring ourselves to sell. We knew what this name meant. It meant a columned, two-story, domed courthouse on the corner of the one traffic light in town, where Main crossed Parker, and the tall cone-shaped cedar on its lawn that was wired with colorful blinking lights at Christmas. It meant Tootle's Bakery where day-old doughnuts were baked into George Washington pudding. It also meant the deteriorating Roxy Theater Daddy had bought, having as a boy watched many a picture show in it, thinking to make a church one day across the street from the jailhouse—the same building we spent one entire summer helping Daddy roof, for which he paid each of us a five dollar bill.

With permission we walked to Shaw's, a living-room grocery neighbors operated about a third mile away, where we traded bottles for a round of soft drinks and honey buns too, if we were so prosperous as to afford both drinks and a gallon of gas.

By the time they got a car running, the boys were barely teenagers, and we were off, traversing ten acres of junk in an old Metropolitan Nash, three Southern kids who talked slow, said *ain't* and *feller,* bumping and laughing around and around the junkyard.

On a Thursday evening in May, when Dell was fourteen and Steve yet twelve, they burned themselves badly on a car radiator. The old car they had been driving around the junkyard was running hot, and when they unscrewed the radiator cap, it sprayed boiling water all over their faces, shoulders, chests, bellies.

Mama was watering flowers when it happened. Dell sprinted across the yard, jerked the hose from her and started spraying himself all over. Mama was screaming "What's wrong? What's wrong?" When there's trouble she panics in an instant and flips into hysterics. The boys were beside themselves with agony and could explain only in bits and pieces of sentences blurted between desperate dousings of cold hose-water. They were in such pain they floundered on the ground, crying and moaning and gritting their teeth.

Mama didn't know what to do. Daddy had ridden off with Mr. John D., who owned the hardware, out to his farm. She hustled the boys to the car and sped to the Lewis farm. Probably the boys should've gone straight to the emergency room, but we had no money and no insurance. More than that, however, Daddy tried to live a life of faith, which relied on miracles of God over those of doctors. God healed the sick and the afflicted. In fact, from the time I arrived out of the universe and landed in

the hands of Dr. Bedingfield, who delivered all four of us,
I never saw a doctor, never took antibiotics or penicillin,
and never missed a day of school for sickness except in
second grade when I had the mumps. Early-childhood
immunizations through the health department ended
our dependence on medicine, until I was in my late twen-
ties, in fact, although a few times we required a dentist,
like the time I knocked out my front teeth playing base-
ball and also for cavities.

So when the boys burned themselves, Daddy, who
keeps his head in time of trouble, telephoned the drug-
gist—it was after hours—who met them at the drugstore
and sold them a can of Foille's spray. Which helped. The
boys' torsos were red as raw meat and blistered like me-
ringue. They could not wear clothes touching their skin
and for a solid week lay wrapped in sheets in their beds of
pain. Even missing school was not worth this.

Away from home we were ashamed of the junkyard. Our
daddy was a junk dealer, but when we filled out his occupation on forms from school we wrote "salesman." We
weren't allowed to socialize outside of school, so class-
mates didn't come home with us, and we didn't go to
other kids' houses. The junkyard, then, was all we knew.
We knew nobody else lived like we did, but we didn't
know how they lived. We knew they were wasteful and
threw perfectly good things in the garbage, which ended
up at our house. We thought that meant they were better
than we were.

The only time as a child I was not ashamed of where
I came from was when a friend from school showed up
with her father, looking for lug nuts or a condenser—or

the time pole vaulters came from my sister's school to tear foam from out of the old seats to land on. Or when the Spirit Club at the high school needed tires for a bonfire. Or when we paraded our Metropolitan through rabbles of townspeople in the homecoming procession.

Then I had something to offer.

It didn't take many years to realize I was a Southerner, a slow, dumb, redneck hick, a hayseed, inbred and racist, come from poverty, condemned to poverty: descendant of Oglethorpe's debtor prisoners. Descendant of people who pulled from the Union, fought their patria, and lost.

When I was in sixth grade my family took one of its only vacations, to Philadelphia for a church convention. We stuck our pinkies in the cool crack of the Liberty Bell, wishing to hear it peal. We tramped wide-eyed through the open air market, blocks and blocks of tangerines, chicken feet, fresh bread, pineapples, fish, cheese, bananas. Glued together, we stared at kids playing in gushing fire hydrants on city summer afternoons, remembering our pond and feeling sorry for them. One day, while shopping for sandwich makings in a neighborhood market, people started crowding close, eyeing us.

Daddy turned to the produce clerk, who wore a brown apron.

"What's going on?" he asked, askance.

"They want to hear your accent," the clerk said. His words were clipped.

I was with one of my brothers by the vegetable cooler when we realized some of the kids had advanced and stood a few feet away. We looked back at them.

"Say something," they giggled.

"Why?" one of us asked.

"We want to hear you talk."

Dell and I smiled at the attention and tried to think of something to say. "What do you want us to say?" Dell finally asked. They set to giggling.

"Why do you talk funny?" one of them ventured.

"Why do *you* talk funny?" Dell shot back.

When I went off to college I struggled bitterly to lose my identity with the junkyard, and my Southernness, starting with the accent. Away from Baxley no one knew my past, and I could accelerate my native tongue and desert its vernacular until I was free. Or thought I was.

One of the first boyfriends I brought home was a Ph.D. botanist whose life revolved around the number of petals emerging from a perianth, whether leaves were hastate or lanceolate, stipulate or pubescent. At Thanksgiving I'd visited his folks' grand ranch house on a sandy-bottom lake in central Florida; they had a sailboat and a marble statue on the lawn. My Daddy, on the other hand, barely finished high school, and neither parent has been west of the Mississippi. I tried to warn the botanist, promising that we might find interesting specimens.

The year I entered high school we moved out of the white house my father built before I was found, into a new house we had erected in one year of hard labor with little outside help. At that point the junkyard officially started at a trapezoid-shaped wire gate that had to be dragged closed, and the junk was contained behind a hogwire fence and manageable. But over the years junk accumulated and much of it didn't make its way into the junkyard. Although unfinished, the new house was two-story, red-brick, nice-looking—a butterfly camped on a dungheap.

Much of the stuff was valuable and worthwhile. An antique washpot lay bottom up near the crabapple tree; a lawnmower that actually worked was jammed under an old truck out of the rain. The rest Daddy held onto simply because he figured he'd need some piece of a piece one day, to fix something, to sell. What classified as junk to most was treasure to him. Mama gave up trying to keep the junk from around the house and instead mowed around the heaps.

Sitting in the front yard was an old fire-gutted, rotted-out airstream trailer, most of its windows busted out. Somebody sold it to Daddy for fifteen or twenty dollars and he hauled it home, thinking to store more junk in it. Eight or ten junk cars got towed in and unhitched in the yard, because the junkyard had spilled over. Making room for the spillover meant delving into the viscera of the yard and consolidating junk, maybe stacking cars on top of each other.

The closer we got to Back Slough, as a friend calls Baxley, the more nervous I got.

"Don't freak when you see the place," I said. "There's junk everywhere. I mean everywhere. Every corner of the house is packed with junk."

Nothing could have prepared him. He vanished to his bedroom and stayed there. He wouldn't come out except for meals, and we left the next day; soon after he called off the relationship. I still think the botanist ditched me because I come from scavengers.

Turning back to embrace the past has been a long, slow lesson not only in self-esteem but in patriotism—pride in homeland, heritage. It has taken a decade to whip the shame, to mispronounce words and shun grammar when mispronunciation and misspeaking are part of my dialect, to own the bad blood. What I come from has made me who I am.

Built by Fire

A couple of million years ago a pine fell in love with a place that belonged to lightning. Flying past, a pine seed saw the open, flat land and grew covetous. The land was veined with runs of water—some bold, some fine as a reed. Seeing it unoccupied, the pine imperiously took root and started to grow there, in the coastal plains of the southern United States, and every day praised its luck. The place was broadly beautiful with clean and plentiful water sources, the sun always within reach. In the afternoons and evenings, thunderstorms lumbered across the land, lashing out rods of lightning that emptied the goatskin clouds; in those times the pine lay low.

The lightning announced itself lightly to the pine one summer evening, "I reign over this land," it said. "You must leave immediately."

"There was nothing here when I came," said the pine.

"I was here," said lightning. "I am always here. I am here more than any other place in the world." The clouds nodded, knowing that lightning spoke true.

In that short time, however, the pine had begun to love the place and called out, "Please. You live in the sky.

Let me have the earth." The clouds glowered and began to thicken.

Lightning was extremely possessive and would not agree to divide.

"Then do what you will," said the pine. For years they warred. The lightning would fling as many as forty million bolts a year at the tree, striking when it could, the pine dodging and ducking. A single thunderstorm might raise thousands of bolts. Wind helped the tree, and although it was struck a few times, the damage was never serious.

After the tree had reached a fair age—old enough for government work, as they say—on the hottest of summer afternoons lightning crept close, hidden by towering maroon thunderheads, and aimed for the tree, sundering its bole crown to roots. When the lightning glanced the ground, such was its ferocity that it dug a trench wide enough to bury a horse before its force subsided. Needles from the pine had fallen about, like a woman's long brown hairs, and they began to smoke and then to flame, and from them fire spread outward, burning easy and slow.

In its dying, the pine sprang forth a mast of cones filled with seeds. The wind played with the seeds and scattered them for miles. And because the mineral soil was laid bare by fire, they could germinate.

But lightning was not worried. Kindling the whole place didn't take much effort. Once lightning struck, the fire might burn slowly through the grasses for weeks, miles at a time, arrested only by rivers, lakes, creeks, and ponds. So if the seeds began to grow, lightning would burn them.

Over the decades the fury and constancy of lightning knew no end—every few years it would burn the place again—and the greenhorn pines learned to lay low, sometimes for five or six years, drilling a taproot farther and farther into the moist earth, surviving the fast-burning, low-intensity fires of lightning's wrath by huddling, covering their terminal buds with a tuft of long needles. Sometimes the buds steamed and crackled inside their bonnets.

Young trees that mimicked grass survived fire. That low, they didn't look like trees.

The grass-trees began to learn that if they waited until the lightning went to sleep in the rainy springs and suddenly cast themselves upward, to the height of a yard or more in one season, drawing nutrient reserves from their long, patient roots, and if they hurriedly thickened the bark of their trunks, a lamination, then when the fires came again they could withstand the heat and their terminal bud would be out of flame's reach.

Only then would the trees dare to branch.

Lightning was nonplussed. No matter what it did, the trees flourished and multiplied. Admiring the courage of the longleaf pine, other trees, hardwoods—sweet gum and sumac and oak—tried to settle. Always, not knowing the secret history of longleaf's adaptation, they burned.

And then lightning realized the pine tree was plugging its needles with volatile resins and oils, rendering them highly flammable. The tree, of course, only thought to make the fires burn rapidly so danger would pass quickly. Flammability was important in driving wildfire through the forest, in order to leave older trees unharmed. The longleaf grew taller, spread farther.

The lightning saw volatility as an act of remuneration.

Longleaf and lightning began to depend on each other and other plants—the ground cover grasses and forbs, or flowering herbs—evolved to survive and welcome fire as well. Wiregrass, for instance, would not reproduce sexually in lightning's absence. The animals learned to expect fire and to adapt. They scrambled off or took cover: down into tortoise burrows, up into tree crowns. During a fire, exotic insects never otherwise seen would scurry from the plates of bark, scooting up the tree. Snakes and tortoises would dash for their holes.

Longleaf became known as the pine that fire built.

Iron Man

My grandfather knew the woods by heart. He was a wild man, Charlie Joe Ray, a ne'er-do-well, born to Scottish immigrants of the Clan McRae, people known even in the Old Country as the "wild McRaes." A group of them had settled in a town in southern Georgia they named McRae, thirty miles from Baxley, and in the even-smaller Scotland, Georgia, nearby. Charlie's father, Joe Alexander, had been a fighter, a champion in the popular wrestling matches that entertained farming immigrants of the coastal plains.

Some say Joe Alexander might have been the strongest man in Wiregrass Georgia in his day. The size of his neck collar was 19½. One day, the story goes, his mule broke into another man's field. The neighbor fetched Joe, and during the ensuing argument, Joe bodily lifted his mule out of the man's patch and heaved it back over the fence.

In 1911, when Joe was thirty-five, he died fighting. He had faced off with a man in the ring in downtown McRae. When the fight was over, Great-Grandpa had won. He staggered toward the town well for a drink of water, and while his back was turned, his opponent weaseled up

from behind and grabbed him in a lock that ruptured
Joe's intestines. He never recovered.

Grandpa was twelve at the time. Soon after, Charlie's
mother, Vana Lavender, remarried, but a year later she
died in childbirth, and Charlie was orphaned, thrown
unwanted into a bastard world. He was taken in by an
uncle. I never heard my grandpa talk about what his childhood was like, except to learn that he struck out on his
own at fourteen. Life at his uncle's had to be hard for
him to abandon the only home he had when but a boy.

Because he withdrew often to the woods for safety and
comfort and for shelter and food, he knew them like nobody I've ever known. All his life he never loved a human
the way he cherished woods; he never gave his heart so
fully as to those peaceful wildland refuges that accepted
without question any and all of their kind. He was more
comfortable in woods than on any street in any town.

But this I heard from his children, my aunts and
uncles: as a father Charlie was poor at providing. He
might bring home two bushels of Irish potatoes and not
be seen again for three weeks. He would expect you to
live off potatoes. "Probably he shouldn't have married
or had children," my father surmised. "He wasn't that
kind of man."

He was terrifying, prone to violent and unmerited
punishment that caught you unawares. For dragging feet
in a bean patch, he might flog a child with a gallberry
switch or the handle of the hoe or a fence rail, anything
he could find. You felt, Daddy said, that he could beat
you to death.

"I tried never to get caught alone with him," he said.

Part of that can be attributed to Charlie's own suffering

as a child and part to the mental illness that came streaking through bloodlines to rest upon his head. Likely he was manic depressive, bipolar, an imbalance defined by flamboyant highs and pitch-black lows. The disorder has run through my father's side of the family for at least three generations that I know of, although kinfolks have said that many of my grandfather's ancestors had been crazy. The illness courses most strongly through the men, although women are not spared, and appears to be caused by a combination of genetic predisposition and stressful environment. Usually it lands for a time upon a person and then may never recur. The illness took my grandfather, and at random it hit among his brothers and sisters, sons and daughters, grandchildren, nieces and nephews. It took my own father.

As a grown man, Grandpa would disappear for days into the floodplain swamps of the Altamaha River, a truly wild place then (even now miraculously unchanged), where he hunted and trapped, fished and plundered. People still remember how he roamed the woods; Charlie was a folk hero.

He could tickle fish. Some people call it cooning, some say noodling. He would leap into the chocolate milk of the alluvial Altamaha and disappear underwater, head and all. He loved to do this when he had an audience—while onlookers waited for a break at the surface, he'd dangle his hands among sunken tree roots beneath cut banks, feeling for fish. When he rose from the murky water, dripping, he hoisted two ancient channel catfish, blue gray as a moonlit night and as whiskered as a circus trainer, one in each hand. He possessed a sort of magic when it came to nature. People were afraid of him.

Charlie hunted coons for a living, by night following his dogs through the south Georgia lowlands, returning by day with his bounty. He might come home with one coon, might come home with thirty. He might not come home. My father tells me Charlie sold the hides to Sears and Roebuck—a beautiful skin could bring ten dollars. The sides of my grandma's house, while he lived there, were stretched tight with curing hides.

Because raccoons are nocturnal, it's dusk-dark when black-and-tan and Walker hounds are set loose in the woods to rush about until they fall upon fresh scent. Then they're on the coon's trail. The hunters are intimate with the meaning of their vociferations, even to the timbre of a certain hound's yipping. The hunters know when the coon has fooled the dogs and they have lost its scent, and they know when the coon has treed and which dog has done the treeing. Through the black timber they make their way to the tree, and either they shoot the animal or they call the dogs off.

The way I understand it, many good hunting dogs open their eyes on this world and excel with adept training, but once in a lifetime, if you're righteously lucky, you raise a hound so aggressive, so fast, so smart, and so dedicated that she redefines hunting. For Grandpa, this was Old Mack. Grandpa traded one of Granny's milch cows for the dog, a move that caused a big rinctum with Granny, who needed the milk for the children. According to Grandpa, Old Mack was special, worth the price of a cow. When the dog ran, he cleared the titi, a lowland bush that betters eight feet, and, in fact, stayed in the air half the time while tracking. Occasionally he touched down to take a few running steps. Old Mack was fast as a bullet, mean as a hornet, as sure as income tax. He could do anything, find anything. You didn't order him to find a coon, you declared how many you wanted and what colors they should be, and he got them for you. He was a thousand-dollar dog.

"Your grandpa was like the Aztecs," my father recounted. "They did a lot of carving in stone, but it was all victories, no defeats." Something happened coon hunting one night that embarrassed Grandpa because it shamed Old Mack and proved him not the dog of prowess Grandpa favored believing. Uncle Nolan, Grandpa's oldest boy, had been along and he's the one who told the story.

The bunch of dogs had treed a coon, or so it seemed from the sound of their baying. When the men gained the dogs, it appeared the coon was not up a tree as expected but encircled by snarling dogs on the ground. The men lit and shined their lights. The biggest boar coon you'll ever lay eyes on—big as a bear cub—was standing on his hind feet in six or eight inches of water, circled by curs. One by one a dog would charge the coon, but before the others could race in, the boar coon had ripped the dog up and hurled him aside. Blood spurted from limp dogs. This raccoon was fierce and mean. Here was the chance for Old Mack to prove himself, not hang back, as he was currently doing.

Grandpa called Old Mack to him and started to rally him in an encouraging voice, scratching him between the ears and caressing his muzzle. Old Mack knew what this meant. He was going in.

Grandpa stood up and sicced the hound. "Git him," he said.

Amid the terrible racket of the baying dogs, Old Mack went for the coon. In a split second, the coon had ripped open Old Mack in a few places, had buried his claws in the dog's head, and was holding it underwater.

Grandpa began to dance and scream, "He's killing Mack! He's drowning him!" He waded in and grabbed the raccoon himself around the neck and began choking him. The men were watching, none of them anxious to tangle with a boar coon. Charlie dragged the coon from the swamp water.

"Here," he said ferociously to Uncle Nolan. "Beat his brains out."

One evening Grandpa was out in the woods with a bunch of men coon hunting. One of the fellows wanted to trade Charlie a hunting dog, and they were testing it. They were on a coon's trail when it treed, but the dogs couldn't find which tree the coon had gone up. Even Old Mack was confused. By chance Grandpa spotted the coon, high in a tulip poplar.

"Stop the dogs," he ordered. *"I'm* gonna tree that coon." The men looked at him as if he was crazy, which he was.

"Stop the dogs," he said. "Old Mack. Git over here."

The men laughed and called the dogs. Grandpa hunkered down on his hands and knees on the forest floor and commenced to howling and hollering, sniffing at the ground. He would run in a short line, sniff around a tree, then run to another. Between trees he spun in circles. Then he ran straight to the tulip poplar where the coon was, stood upright and pawed at the damp bole, baying.

The men lit the lightered knots, pieces of the resinous and volatile wood of aged longleaf that burns as hot and bright as lighter fluid, they carried and thrust them high. Sure enough, raccoon's yellow-green eyes glittered from an upper branch, its ringed tail blurry against the bark in the tattered light. The men stared at Grandpa with their mouths hung open and an uneasiness sidled through the woods.

"How did you know where that coon was?" one of them demanded. "You watched him tree?"

"I smelled him," Grandpa said. "Come here. Git right down here and you can smell him too. His scent goes right through here, this way, then he heads straight to that big tree." He was on all fours again, demonstrating.

None of the men offered to go down and smell the ground.

A joke is only a joke if there's an end to it, when everyone gets "let in," and Charlie wouldn't let anyone in. People's wariness of him was strong, but he seemed to shrug off friendship, and honesty's part in it. Until he got too old to tell stories, he related his trickery to us, his eyes bright and glinting, his huge hands reenacting scenes of craftiness in the flatwoods.

On another hunt Grandpa set the woods afire with fat lightered. A coon had disappeared into a hollow tree streaked with heart pine, the heavy, resin-saturated heartwood, and Grandpa proceeded to build a fire in the hollow to smoke the animal out. Before long the fire was blazing in the tree trunk, melting resin and steadily climbing until it leaped high into the chimney of tree. The tree was seventy-five feet tall and hollow clear to its crown, and soon hammering and hissing noises issued from it as the fire clambered skyward.

"Be damned," someone drawled. "The whole thing's on fire."

According to whoever told the story, fire blew out the top of the tree and flared, a wild and extravagant orange flower in the swamp. Below, hunters were whooping and running through the woods. Fire shot one hundred feet into the air above the tree like a blowtorch, burning a half-mile of pinewoods that night.

When their first children were small, Grandpa Charlie and Granny Ray, whose name was Clyo, started a restaurant in town. It was the late 1920s, maybe 1929, and Daddy wasn't born yet. Nobody remembers the name of the cafe. Somebody loaned Grandpa the storefront free, over on Comas Street, and with $1.69, Grandpa bought a frying pan, a loaf of bread, and some hamburger meat, then went looking for customers. Over the months he and Granny built up a satisfactory business, but even after people were coming in regular, Grandpa would take an order and then have to zip down the alley to the store to buy what the folks wanted.

Two terrible fights happened at that restaurant. One was the time Grandpa almost killed Isom Copeland. Copeland was a neighbor to Grandpa and Granny, with only Ab Turner's strip of land separating them. How the fight happened was that somebody stole a five-gallon jimmyjohn of whiskey Copeland hid in the woods. Anyone rambling could have taken it, and Copeland had teenage sons at the time who knew where it lay, but he suspected Grandpa.

Isom rode up to the restaurant on a white mule and yelled for Charlie. Grandpa came out holding my uncle Nolan, a small boy, in his arms. Right on the street Copeland, highly angered, accused Grandpa of theft and hit him across the face with the jaws of his knife.

Grandpa set down Nolan. It is said that he would not start a fight but he would end it. He fought willingly, and when he fought, he fought to kill. Like his father, Joe Alexander, Charlie was abnormally strong, built like a barrel, his fingers so massive he couldn't buy a wedding ring.

Grandpa like to broke the man's neck. When he finished with Copeland, he doubled the man across his wild-eyed mule, which had come up white but left red. Grandpa slapped the mule's rump, and it trotted toward home with its mangled burden. Shotgun in hand, Grandpa lit out for the woods without so much as a fare-you-well, where he hid out on the lam for a few days, but nothing ever came of the fight, as they say, meaning Grandpa wasn't charged. As for the other man, the rest of his life Isom Copeland trod looking at the sky, unable to lower his head.

The cafe lasted a couple of years and ended with another fierce fight. This time a boxer had come to town and inauspiciously happened to choose Grandpa's cafe. It was a Saturday morning.

"Morning," Grandpa said, squinting against the new bright sun at the stranger.

"Morning to you," the man replied. He was wearing shined shoes and carried a satchel. He found a seat. The restaurant was quiet with the men who came around every morning to drink a cup of Granny's black chicory coffee and hear what new had happened since the sun last set.

Grandpa asked the man if he wanted coffee and he said, *of course.* There was something cocky and abrupt Grandpa didn't like about him. Granny Ray stood at the sink scrubbing a pot. She heard the man and poured a white cup full of coffee.

"I'll see to him," Granny said. "You finish frying Horse's bacon."

"Naw," replied Grandpa. "I'll take care of him." Granny looked Grandpa straight in the eyes and handed him the cup. When he took the crockery mug it disappeared inside his fist. He handed it to the stranger.

"Your face is new around these parts, ain't it?" said Grandpa.

"Yessir," said the man. "I'm a boxer. I came here to Baxley to fight. I've heard you've got some good fighters here, and I'm gonna take them on." At that time yeomen thronged Baxley's active fighting rink, a rowdy place on Saturday nights.

"You fancy yourself a boxer," Grandpa said.

"Best there is," the stranger boasted.

"I'm a boxer myself." Sometimes Grandpa went down and fought in the rink, when somebody would contest him.

"I've fought up and down the East Coast," said the man. "Never been beat." That galled Charlie, but he said nothing.

"What'll you be having to eat this morning?"

The stranger ordered grits, an egg, and biscuits. "You fry my egg in butter," he said. Grandpa was bristling. "I have to have it like that, to stay in shape."

"I fry mine in grease," said Grandpa. "And I'd whip yo ass anytime."

"Let me see you do it," said the man, jumping from the table, upsetting his straight-leg chair, which was hided with one of Grandpa's cow skins, the hair still on. Overturned chairs don't bode well. Grandpa and the stranger faced off.

"Charlie, what are you doing?" Granny came running from the back. She picked up the chair as if that would stop the men who were dancing lightly, swaying their upper bodies from side to side. They looked like disturbed rattlesnakes.

"Get out of here right now," Granny yelled. She stepped up to their double knot. That's when the boxer threw his first lick, an action that turned out to be the biggest mistake of his life. It was a fight to be remembered. Within seconds the place was vacated, and the battling men went at each other's throats like feral dogs.

Grannny was outside the door yelling, "Stop them, somebody stop them! They're going to kill each other!" but no one moved a muscle to wade amongst the fracas or even to go for the law.

In their rage the men forgot who and where they were. Nothing could stop their blind desire. They overturned tables and broke chairs and splattered blood on the walls. Even after the stranger was down and unconscious, onlookers had to pull Grandpa off him. He stood up, still enraged—his face and neck engorged with blood, and visible through torn clothes the ropes of muscles that strapped the magnificent machine of his body—and stumbled to the doorway. He lowered himself and sat for a minute or two on the step. There was silence the world over. Charlie spat and a clump of blood swirled and marbled in the dirt. He raked his thick hair back and would not look at the people gathered in the street, now easing away from him. Then he suddenly stood up, grabbed the traveler's satchel and crossed the restaurant to where the man lay unconscious, battered cheek against the splintered wood floor. He didn't look so good.

Grandpa picked him up, folded him over his shoulder, and stepped into the street. He turned south, leaving footprints deeply embedded in the fine street dust. The railroad depot was only half a block away, and though with every step Charlie felt the weight of his bloody load, he walked straight and sure. He neared the ticket counter and spoke through its small window to the astonished clerk.

"This man decided to ride on," he said. He didn't wait but strode toward the open door of the passenger car, where a few people prepared to board. They drew back. Grandpa climbed the steps and dumped the man in the first seat he came to, propping him against the grimy window, then dropped the brown satchel beside him. The man was still out cold, sagged against the train, his clothes torn and dirty and freshly bloodstained.

Charlie ducked into the bright sunlight and into the hush. One of the passengers was a woman wearing cotton stockings and a hat adorned with white egret plumes. She was with an older man whose polished watch fob shone bright gold against the charcoal of his pants.

"Morning," Grandpa said to the woman as he stepped past, as if it were any normal morning in Baxley. He did not return to the restaurant—there was no restaurant to return to, destroyed as it was. Instead, he fetched his dog from home and struck out for the river swamp. And that was the last day of business.

The Copeland incident, however, was not over. For years the sons had nurtured the antipathy of their father. Years later one of them, Shutty, tried to avenge his father's disabling. Grandpa had ridden down to Alma with some fellows to carry a load of pigs, not knowing Shutty would be along. Charlie was at least fifty years old at the time, Shutty twenty years his junior, and big, tall. Shutty misread his advantages.

Through the unloading of the pigs Grandpa felt the young man's tension, which is why, when the truck cranked up to head back to Baxley, he bid to ride in the back. He reckoned Shutty would ride in the cab with the others.

"I'm riding back there with Charlie Ray," Shutty said. Not a muscle moved in Grandpa's broad sepia face. In fact, he pretended not to hear. The other two men glanced at each other, then away toward the long empty highway.

"Shutty," one of them ventured. "Ride on up here with us. They's room."

"Thank you just the same," said Shutty. "I've got business to settle."

Grandpa and Shutty lifted themselves into the stinking, high-sided back and the metal gate lock clanged behind them. It was twenty miles to Baxley.

When the truck pulled into the city limits, Shutty was unconscious.

What I know about Grandpa's mental illness is confined to hearsay. Daddy was yet an unschooled boy when Grandpa was first sent to the state hospital at Milledgeville. He thought my grandmother was cheating on him. Until then, she had been the only person who could control him, who he'd listen to—she stood her ground with him and he respected her. Or perhaps she knew the reach of his love protected her from his wrath.

"Five minutes with her," my grandfather said, "was better than five hours with any other woman."

When he was sick it was a different story. He scared her. He'd make pallets on the floor of the shed and accuse her of bringing a lover there. Sometimes he thought the "striped wagon gang," whatever that was, was coming after him.

Always strong-willed and independent, Granny Ray signed for him to be sent off. In those days, it took only one family member to sign papers to have someone committed to a mental institution. Charlie was gone for months, but one day he turned up Brobston Street and sauntered to the house. It was evening and Granny had come home late from working at the All-American Cafe. There was a strange air about her house, she thought as she approached. Maybe it was the children. They weren't playing and fussing in the yard as they normally did. Sometimes her youngest sons, Johnny and Little Gene, so called to distinguish him from Jean Tyre, who Nolan, the oldest son, had married, met her at the end of the street and helped her tote what packages she might have.

"Frank?" she called. Daddy was a couple of years older than Johnny. "Where are you? Gene."

It was her husband who answered. He was sitting in the shadows of the porch.

"Try again," he said.

"What're you doing here?" she asked.

"You tell me," he said. She sat heavily on a chair. Her legs were swollen and her feet hurt.

"How'd you get here?"

"I walked," he said.

"All the way from up there?" She didn't believe him. Yet the sheriff had not notified her that he'd be released. He must have escaped.

"How long ago did you leave?" she asked.

"Never mind," he said. Honestly he did not know. Maybe for two months he had been making his way home through the river floodplain, living off the land. Granny wondered if the hospital guards had chased him, and if they'd turned dogs loose to find him.

"I figured you were the one put me in the loony bin. That right?"

"Charlie, I can't go through this any more. You need help," she said. She could see he still wasn't well.

"Help, hell," he said. "I need you to act like a damn wife's supposed to act." It wouldn't do to make him any madder.

"You can't stay here," she said.

"This is where I live," he answered. The children were hovering around the edges of the porch, bug-eyed and interested but not daring to approach too close. Granny heaved out of her chair.

"Let me get supper going," she said. Grandpa spotted Franklin lurching around the corner of the porch. Someone had pushed him from behind.

"Frank!" Grandpa hollered. "Git over here." Franklin jumped back behind the house wall and sprinted for the backyard where he could gain the tree line unhindered.

"If I catch you, you'll wish to hell you'd listened,"
Grandpa hollered.

"Charlie," Granny warned.

"They don't show one ounce of respect. I'll beat sense
into them if I have to," he said.

"Over my dead body," she muttered.

Through the cooking and serving and partaking
of Granny's simple fare, Grandpa rambled about the
house, hounding and berating his tired wife. His anger
she deftly deflected so as to defuse it. The younger boys
stayed out until she called that supper was ready, and
as soon as they ate, she sent them to bed in the room
they shared. Cleo, the second daughter, whom they
called Mutt, arrived home. She was startled to see her
father and gave him a quick and tentative hug, which
served no purpose in softening him. He was beyond affection, she soon learned. Mutt also worked at the All-
American, but her shift kept her late, and the owner's
young red-haired son, who did not stay, escorted her
home.

"So your mama sent me off," Charlie said. "But I
showed her. She might have them put me away, but they
can't keep me."

As the night wore on, Granny's reserves of tolerance
depleted. The yellow moon had come up full over the
scrim of darkly green trees bordering the yard when she
began begging him to leave.

"If I had money I'd pay for you to go from here," she
said, her voice screeching "and let us be. You're keeping
us all up with your arguing. We need sleep. Get out of
here and stay gone." A candle moth beat itself against the
bare lightbulb.

"If I had the money, I'd go this minute," he raged. "The sun wouldn't rise on my face in Baxley, Georgia."

Cleo had just been paid. She and Granny each made fifteen dollars a week.

"I'll give him the money," she said. And did. She gave him everything she had.

That was the night Charlie took off for good, toward the orange groves of Florida, abandoning my grandmother to raise eight children alone. Occasionally he sent letters from Wauchula or Bowling Green, and periodically he roared back into town. He started a new braggadocio life down in Florida, calling himself "Iron Man" and setting records picking oranges, out-picking any who dared challenge him. Once Daddy rode the train to see him. When he arrived, the station was crowded but he saw his father across the throng. Charlie had dyed his hair, which had begun to fade to gray when he was five or six and had been shock-white since he was twenty-one, utter black. He wore a red kerchief knotted around his neck.

"Daddy," my father yelled over the hullabaloo. "Over here."

Grandpa picked his way through the fray. "How'd you know me?" he said.

"I'd know you anywhere," Daddy replied. "You're my daddy."

For portions of my childhood, Charlie returned to live in Appling County, although really he was as absent in his grandchildren's lives as he had been in his children's. He fascinated me, since I had nothing to forgive. He'd show up in his old Biscayne with iris bulbs dug from the swamp or a mess of bluegill for Mama to cook. He was full of jokes and folk songs and hunting stories, his bottom lip tamped with snuff, accompanied by an old dog or two that traveled on the seat beside him.

Unless caught unawares, Grandpa affected a dialect most non-Southerners think of as *the* Southern accent, but which is usually upper-class speech, one of many dialects. It pronounces an *a* instead of *er* at the ends of words like *mother* and *father*. Motha, fatha, the way Jimmy Carter talks. Depending on geography, lower-class speech is more harsh and ground out, often grammatically incorrect, and hardens the ends of words: *yellow* is *yeller.*

"What cula was George Washington's old gray mule?" Charlie would ask.

But we'd heard the joke before from him and knew its answer by now. So he'd impress us by sketching likenesses of foxes and bears, anatomically correct, in our composition books. He drew remarkably well, short quick strokes of pencil lines that did not connect.

We'd eat a meal of hush puppies and grits, and before he left, he'd tell us some long-winded and high-handed tale of his exploits and the great calamities that had befallen him—alligators chasing him and him wrestling them off, or trains bearing down while he tried to rescue his hunting dog whose leg was hung in a crosstie. At the apex of greatest danger Grandpa would stop talking. We'd stare dumbfounded, awaiting the denouement.

"Fish haven't been biting atall lately," Grandpa would change the subject. "I've been down to the river every day this week and haven't caught anything except one little ole stumpknocker."

"What happened, Grandpa?" we'd want to know.

"Nothing," he says. "I sit for a solid hour at Eason's Bluff. I was fishing with crickets. Maybe I should've been using catalpa worms. They're in the trees now."

"No, no," we'd protest. "What happened with the train?"

"Well, it hit," he'd say.

"Did you die?"

"Now how would I be dead?" he'd say. "I didn't say the train hit me. All it hit was an old stick I wedged in the dog's mouth, to keep her from biting, cause I left it behind, getting out of the way. Nobody got hit. I had time to pick up an old Co-Cola cap and put it on the track. Still got it, in fact. Here, who wants a lucky bottle cap your Grandpa's been toting in his pocket for seventeen years now?"

He'd no more carried a flattened bottle cap around for seventeen years than the cow jumped over the moon, but we didn't know that. Charlie loved to hear himself talk, spinning yarns of greater and greater incredulity. Sometimes he'd talk into a tape recorder so we could listen to it. He could fill one side of a tape with political ranting, about the Puerto Rican migrant workers taking over Florida orange groves or preaching and the flipside with dirty ballads. We'd beg him to sing and he would, in a deep, resonant voice. I've forgotten all the bawdy songs.

You get a line and I'll get a pole, honey.
You get a line and I'll get a pole, babe.
You get a line and I'll get a pole,
We'll go fishin' in the crawdad hole, honey, baby, mine.

Charlie dipped Navy snuff kept in a tin in his shirt pocket. To take some he pulled out his lower lip and tilted the can to it. When Grandpa was a boy, his sister

Rosalee had miscalculated the aim of her ax while split-
ting kindling that Charlie was holding, and she had
chopped off the middle finger of his left hand. A nail had
grown back on the trimmed finger but it was round as a
toothpick, as big as a match stem. Grandpa used that nail
to tap snuff into his lip and it pinged against the can,
ping, ping, as the powdery brown tobacco tumbled in.

One day while visiting us he forgot his snuff down at
the shop and sent Dell to retrieve it. When Dell got back,
the can was empty.

"What happened to my snuff?" Grandpa demanded.

"I poured it out, Grandpa," Dell told him innocently.
"It wasn't no good." He had stopped midway to smell
the contents of the can and, finding it dusty and rotten,
dumped it on the ground. Grandpa was livid and would've
whipped Dell, but Daddy's interdict against him punish-
ing us kept his anger half-contained. All Grandpa could
do was cuss and grumble.

"Where's your brains, boy? You poured out my good
snuff, all I had." Daddy didn't want him cussing around
us either.

Sometimes we visited Grandpa in ramshackle shanties—strewn with piles of clothes and newspapers, tackle
boxes and garden tools—he rented for forty or fifty dollars
a month. He lived off a government pension. He always
managed to have a bed, heaped with dingy, pungent, and
moth-chewed blankets, and a few straight-back chairs with
cowhide bottoms. Once, when Grandpa was staying way
out Highway 144 toward Glennville, we went to help him
move to another shack next to Altamaha School, where he
lived until a possum fell into his well and ruined the water.
Daddy was in the almost empty kitchen.

"Daddy," he called. "You've got a snake out here."

We rushed to take a look. Daddy made us stay back.

"Where's he at?" Grandpa said.

"Behind there." He pointed to the refrigerator.

Daddy said the snake, coiled up, would fill a hubcap. "If you butchered him out you could eat for three months," he said. It was dim behind the refrigerator, with the electricity already turned off in the house, and he couldn't tell what kind of snake it was. It looked like a rattlesnake.

Grandpa peered around, kind of grunted, said "I reckon it is," and ambled back into the front room with no more concern than a mailbox.

Grandpa paid us to catch shiners and fish to stock his favorite ponds. A grid of canals had been dug throughout Baxley to drain the town, which had been built in a swamp and was low enough that at times motorboats have traveled Main Street, flinging water against the glass-fronted stores. One canal that ran behind the junkyard seemed to go for miles. The scars of its construction were softened by the wear of time, canal sides no longer perpendicular, rushes and cattails returned, layers of leaves carpeting the mountain range of clay leavings piled high on one side. The canal ran through the woods, where pine needles in their falling forked against the branches of wax myrtles and draped from them like fringes. We could have followed the canal into the deepest wilderness.

I carried an anthology of poetry to the canal and read Tennyson and Wordsworth while my brothers fished. I could look out across the field of weeds, back toward the road, and see vales of heather. There is one particular poem attached to that place, but I cannot recall even one line. When I read it again I will know it.

Sometimes I fished, but only those who dug earth-worms got to use them. It was a tedious job going here to there, lifting rotting boards and car hoods to get at soft black loam riddled with annelid activity. It required a pitchfork because a shovel cut too many of the worms in half, a nauseating and sorrowful thing, even if earth-worms had magical powers of regeneration, as we were erroneously told, sprouting from each end and making two of themselves.

Stringing them on a hook was bad. They writhed and oozed ceolomic fluid into your hands, until your fingers were sticky and black.

Grandpa taught me the basics of fishing, including how to cast a lure. We would fling and reel a gaudy Rapala up and down the drive, until it was second nature to press while casting, then click to set. All three of us had a tackle box and learned from him how to thread bobbers and tie hooks.

Whatever shiners, redeyes, and sunfish the boys caught, they would stow in a bucket for Grandpa, who paid them a dime each, until they got overambitious and netted from the canal more fry than he could afford.

"Whoa," he said. "You boys been talking to Jesus? You know he took one fish and fed a couple of thousand people."

"We've been using a net, Grandpa," Steve said. Grandpa called him Honorable Stephens.

"Better quit," Grandpa said. "I cain't take this many fish. You boys gone break me."

Grandpa also took very seriously my being able to fight, and he would bring me to the grass under the catalpa tree for lessons—how to tighten the elbow against the neck, how to wrap a leg behind, how to throw weight.

"I taught Lee to fight," he said, "and she'll beat hell outa anybody messes with her." Aunt Lee, like me, was one of his favorites.

Grandpa'd have my brothers and me scrapping in the dirt.

"Bring him down! Bring him down!" he'd coach, dancing excitedly around us. "Pull out! Take him down! Pin him, pin him now!" According to his sensibilities, brute power was the most noble of attributes. He loved to see us pitting our strengths against each other. Afterward he would take out a block of hogshead cheese or a muskmelon and carve us slices.

As we got older we didn't topple so often into his tricks and jokes and commenced to start testing him. Like the time he was watching us for a few hours.

"Grandpa, do you have a girlfriend?" A number of old women he called upon frequently, bringing a bunch of collards or a peck of fresh-dug sweet potatoes, insinuation that he hankered after a good meal, and women, being wont to enjoy feeding a hungry man, fed him. Thus he made his rounds from widow to widow, meal to meal.

"A man 'thout a girlfriend's practical worthless," he said.

"Who is it then?" we challenged. We had never seen him with a woman, although his marriage to Granny documented that he had liked them.

"I cain't tell all my bidness," he said.

"You're making it up," we said. "You ain't got a girlfriend."

"I bet you a hunnerd dollars I do."

"Who is she then?"

Finally he allowed it was Essie Mae Higginbotham, who lived out Highway 144 toward Glennville with her brother. We still weren't satisfied.

"Call her up then," we demanded. We wanted to hear what he said to her.

He agreed to call. We looked up the number for him in the telephone book. Higginbotham. He angled the phone away from his ear so that, clustered around the receiver, Dell closest, we might hear some of the conversation.

"Hello," she said sweetly.

"Essie Mae, this is Charlie," he said.

"Beg pardon," she said. "Who did you say?"

"Charlie," he repeated.

"My hearing's about shot," she apologized. "Who?"

"Charlie," he shouted. "Charlie Ray."

"OK," she said. "How you been doing?"

On our end Grandpa bowed up like a rooster. He stabbed at small talk, the heat, the rain, then he cocked his bowstring and shot his rusty arrow of love.

"Essie, you're my girlfriend, ain't you?"

"I'm sorry. What did you say?" This forced him to repeat his embarrassing question.

"No, I am not!" she exclaimed. "Charlie, have you lost your mind? Whatever made you think that? I have as much need of a boyfriend as I do smallpox." And she hung the telephone into its cradle.

Devilishly we laughed at him, hooting and rolling

against each other and onto the furniture in our hilarity, slapping high fives. We were not ignorant of the fact that deriding Grandpa was perilous territory we might be fleeing, as Uncle CJ had fled when he first defeated Grandpa in a footrace. The grown son challenged his father to race in a tobacco patch.

"To the end of the field," Charlie said.

Someone counted and yelled, "Go!" and father and son leapt into the hot, gray dirt of two rows packed by the feet of croppers and the narrow tires of Farmall A's. Charlie pulled into the lead, running hard and dragging air, with CJ at his heels. At the very end CJ edged him and crossed into the selvage of coffeeweed and jimson a half-pace ahead.

If CJ had won silently, the dethroning of the champion would have passed as such events usually do when sons overtake fathers in strength, but CJ would give Grandpa his own vainglorious medicine. He vaunted and crowed and staggered. Charlie hated to be outdone. With bloodlust in his eyes, he swiveled and charged for his son, who sprang away, racing through the furrowed rows where he could not gain ground unless he zigzagged through the tobacco, breaking stalks in his path. Now CJ really had to outrun his daddy.

In his aging Grandpa had become tormentable.

Once, just once, Grandpa led us to a soft, giving place. All of us went. It was a secret spot only Grandpa knew, far beyond pine woods to a sandhill where wild huckleberries grew in deep blue profusion.

We drove to the other end of the county, parked the car on the dirt road, and entered a noisy upland thicket, pine needles complaining of our weight even as they yielded, wax myrtle whipping our bare legs. A far piece we walked through air inundated with the long-summer smell of pine, alcoholic and comforting, until we came to a gabble of shrubs thick with purple berries. After spreading bedsheets underneath, Grandpa beat from the *Vaccinium* bushes a thick and succulent carpet. We drew a washtub of huckleberries from that coppice, so many that we had to rinse them in the bathtub, picking leaves and twigs that floated out. They turned into sweet and seedy cobbler, which ran purple channels across flour dumplings on our plates, and into huckleberry pie.

Charlie could've tutored me, had he been able, in the swamp's secrets—how to survive as an orphan there and how to survive in general. Even as a young girl I desired this unwrought knowledge and knew it unreachable, fenced as it was by Grandpa's anguish, which had driven him toward the solace of the wild in the first place and hindered as it was by Daddy's canon, which restricted daughters to the household and made them mistresses of domesticity and which prohibited an intemperate tramping about. *Of what use to humanity,* I ask myself, *is a man who cannot see beyond his own hurt?*

Grandpa Charlie had a profound effect on me, and often I understand, when I'm in the woods, that I am looking for his secret copse of huckleberries. Something in me seeks the pure amazement I had that day as a girl: the sweet-faced wonder of enough berries to feed Appling County, a plenitude advertised by our purple-stained mouths, amid a beauty too reckless for return. What is left of this mythic terra incognita is a map I cannot follow. I have not stopped trying to go back.

Forest Beloved

Maybe a vision of the original longleaf pine flatwoods
has been endowed to me through genes, because I seem
to remember their endlessness. I seem to recollect
when these coastal plains were one big, brown-and-tan,
daybreak-to-dark longleaf forest. It was a monotony one
learned to love, for this is a place that, like a friend, offers
multiplied loyalty with the passing years. A forest never
tells its secrets but reveals them slowly over time, and a
longleaf forest is full of secrets.

I know a few of them.

Longleaf pines are long-lived, reaching ages over five
hundred years. As trees mature, their heartwood becomes
so heavy and thick with resins that saw blades bounce
away from it, and if saw teeth manage to enter the grain,
they quickly gum up and dull. Heartwood mills a strong
everlasting timber the color of ripe amber that earned
longleaf the name "heart pine." Parcels of the tree, especially stumps and the area of the heart, are more heavily
tamped with resin, and that wood is called "fat lightered,"
though people use the two names interchangeably, "heart
pine" and "fat lightered," and sometimes they say only

"fat," as in "Put another piece of that fat on the fire." It is so rich with concentrated, cured sap that it burns like a flare and has long been used, in very small pieces, as kindling; the resinous knots were early lanterns.

In the heart rests both the tree's strength and its weakness. After about ninety years, pines often are infected with red heart, a nonmortal fungus that makes the heartwood softer, more porous, and more flammable, and that often hollows out the pine and makes of it a refuge.

You don't think about diversity when you look at longleaf. In a fully functioning longleaf woodland, tree diversity is low. A single species of pine reigns in an open monologue of tall timbers (except on sandhills where occurs an understory of turkey, post, and bluejack oak). The trees grow spaced so far apart in pine savannas, sunshine bathing the ground, that you can see forever; they are as much grassland as forest. The limbs of longleaf pine are gray and scaly and drape as the tree matures, and its needles are very long, up to seventeen inches, like a piano player's fingers, and held upright at ends of the limbs, like a bride holds her bouquet. In 1791, naturalist and explorer William Bartram, in his *Travels,* called the Southern pinelands a "vast forest of the most stately pine trees that can be imagined."

The ground cover, a comforter laid on the land, contains the diversity. *Wiregrass* dominates—it's a flammable, thin-leaved, yellowish bunchgrass that grows calf-high and so thick it resembles a mop head. From this sinewy matrix of wiregrass all manner of forbs, grasses, and low shrubs poke up. At every step, another leaf shape or petal form begs examination and documentation.

Meadow beauty. Liatris. Greeneyes. Summer farewell. Bracken fern. Golden aster. Sandhill blazing star. Goat's rue. Yellow-eyed grass. Purple balduina. Beautiful pawpaw. Pineland hoary pea. Wireleaf dropseed. Hair grass. Little bluestem. Lopside Indian grass. Toothache grass. Britton's beargrass. Gopher apple. Dwarf live oak. Low-bush blueberry. Blackberry. Runner oaks. Splitbeard bluestem. Honeycomb head. Croton. Clammey weed. Dog tongue. Rayless goldenrod. Narrow-leaf sunflower. Black-eyed Susan. Dwarf wax myrtle. New Jersey tea. Inkberry. Dwarf chinquapin. Cooley's meadow-rue. Chaffseed. Sandhills milkvetch. Spurge ipecac. Wireweed. Sandwort. Blue lupine. Winter-flowering pixie-moss. Clasping warea. Pigeon wing. Toothed savory. Hairy wild indigo.

One hundred ninety-one species of rare vascular plants are associated with longleaf/wiregrass, 122 of these endangered or threatened.

When John Muir conducted what he termed his "floral pilgrimage" to the Gulf in 1867, somewhere on the fall line between Thomson and Augusta, Georgia, he decribed "the northern limit of the remarkable longleafed pine, a tree from sixty to seventy feet in height, from twenty to thirty inches in diameter, with leaves ten to fifteen inches long, in dense radiant masses at the ends of the naked branches.

"The wood is strong, hard and very resinous," he wrote. "It makes excellent ship spars, bridge timbers, and flooring." Later he added, "I thank the Lord with all my heart for his goodness in granting me admission to this magnificent realm."

What thrills me most about longleaf forests is how the pine trees sing. The horizontal limbs of flattened crowns hold the wind as if they are vessels, singing bowls, and air stirs in them like a whistling kettle. I lie in thick grasses covered with sun and listen to the music made there. This music cannot be heard anywhere else on the earth.

Rustle, whisper, shiver, whinny. Aria, chorus, ballad, chant. Lullaby. In the choirs of the original groves, the music must have resounded for hundreds of miles in a single note of rise and fall, lift and wane, and stirred the red-cockaded woodpeckers nesting in the hearts of these pines, where I also nest, child of soft heart. Now we strain to hear the music; anachronous, it has an edge. It falters, a great tongue chopped in pieces.

Something happens to you in an old-growth forest. At first you are curious to see the tremendous girth and height of the trees, and you sally forth, eager. You start to saunter, then amble, slower and slower, first like a fox and then an armadillo and then a tortoise, until you are trudging at the pace of an earthworm, and then even slower, the pace of a sassafras leaf's turning. The blood begins to languish in your veins, until you think it has turned to sap. You hanker to touch the trees and embrace them and lean your face against their bark, and you do. You smell them. You look up at leaves so high their shapes are beyond focus, into far branches with circumferences as thick as most trees.

Every limb of your body becomes weighted, and you have to prop yourself up. There's this strange current of energy running skyward, like a thousand tiny bells tied to your capillaries, ringing with your heartbeat. You sit and lean against one trunk—it's like leaning against a house or a mountain. The trunk is your spine, the nerve centers reaching into other worlds, below ground and above. You stand and press your body into the ancestral and enduring, arms wide, and your fingers do not touch. You wonder how big the unseen gap.

If you stay in one place too long, you know you'll root.

I drink old-growth forest in like water. This is the homeland that built us. Here I walk shoulder to shoulder with history—my history. I am in the presence of something ancient and venerable, perhaps of time itself, its unhurried passing marked by immensity and stolidity, each year purged by fire, cinched by a ring. Here mortality's roving hands grapple with air. I can see my place as human in a natural order more grand, whole, and functional than I've ever witnessed, and I am humbled, not frightened, by it. Comforted. It is as if a round table springs up in the cathedral of pines and God graciously pulls out a chair for me, and I no longer have to worry about what happens to souls.

Junkyard

When my father was eighteen, Charlie, in one of his so-journs, talked his son into a partnership in a wrecking yard business.

"There's money to be had," he said. And there was.

In 1955, the junkyard began as a chrome glint in my father's eye. Over five million motor vehicles were produced in U.S. plants that year, according to *Auto Facts and Figures,* and two out of every three passenger cars and three out of every four trucks manufactured were sold as replacements for vehicles scrapped. My father got his share. In those days, especially in the rural South, most people were poor enough and handy enough to fix their own cars. They ransacked junkyards for replacement parts—voltage regulators, alternators, windshields, drive trains, steering sections, hubcaps—in lieu of expensive new ones from the automotive store. If the generator of a '58 Chrysler went bad, maybe one from a '57 or '59 might work.

Daddy bought or traded for junk cars people had lying around their places, paying them five or ten dollars for "the favor of hauling it off." If the cars were on blocks, he stuck tire axles underneath and towed them away. If they had tires, he aired them up and changed what needed changing.

Many of the cars immediately sold for scrap. With a hammer, chisel, and ax, Daddy and Grandpa cut the frames loose from the chassis in order to separate iron, since tin was bringing a cheap price. They hauled the metal to Chatham Iron and Steel in Savannah, where it was loaded in boxcars. Before long, however, father and son had come to a disagreement. Grandpa desired nothing other than to earn enough money to live week to week, but my father wanted more, an inventory in order to advance. So he bought out Grandpa's share in the junkyard and went into business with an older entrepreneur, Lee Woods, for a couple of years before buying him out as well.

He was a great salesman, my father, but he was the greatest buyer in all the world. He could get anything he wanted for next to nothing. Psychology was the trick. Buying something, you had to be patient, never overeager to own a piece of merchandise. You had to act like owning it was neither here nor there.

"You want to get rid of that old jeep over there?"

"Well, we kinda want to fix it up one day."

"It's about rusted out, ain't it? How long's it been sitting there?"

"About five year. I inherited that from my brother when he got killed in the car accident, and I'd hate to part with it."

"I remember hearing about that," my father would say. "Happened out Ten Mile Road, didn't it?" My father's hair is prematurely gray like his father's. His speech is full of passion and feeling, his manner one of conviction, and he is deftly charming.

"Yeah, a tractor trailer crossed the center line head on."

"I shore remember it," and without change of expression, my father would skillfully switch the subject of conversation, usually to something close to the owner's heart, maybe a penned hunting dog or a towheaded child dragging a stick around.

"My daddy used to keep chickens just like those. Dominickers, he called them."

From there the conversation would meander, touching on anything in the world except the jeep. Daddy would inquire idly about the functionality or availability of other sundry items thrown around, a busted harrow or a motorcyle frame, a stack of old windows or a heap of tobacco sticks, but nine times out of ten, even if the original answer had been an immediate and adamant *no*, he towed the jeep home. Sometimes he left without it but got a call within a few days, the people saying they'd changed their mind.

He never rubbed a bad deal in someone's face, meaning if he had the jeep running in two hour's time and tripled his money when he sold it, he didn't call back to announce his good fortune. He let the sellers think they'd bargained well and won.

I've steered many a car home for my father, crawling among dust, spiderwebs, and wasp nests onto crumbling upholstery, hugging the wheel while we lurched and jerked along. If the car had brakes, my job was to keep from ramming into Daddy's truck, which meant us creeping along, me paying close attention to his brake lights and turn signals. I'd be connected to his truck by an eight-foot range of iron chain, a margin of error no longer than a squeal.

If the car had no brakes, we'd rope a rubber tire or two between us—sometimes I'd be jerking the chain, sometimes riding his bumper. The most powerless feeling in the world is to pump at dry brakes while ramming the tail end of your Daddy's International. Itchy flecks of vinyl and foam stuck to my thighs, under the skirt, and sweat popped out like BBs along my ribs.

"Mr. Frank, you got a radiator fit a Ford truck? I got a hole in mine." It'd be a customer, looking for a part. I'd look up from my doll-playing or later from the wee garden where I grew okra and marigolds, hauling buckets of sheep manure to fertilize them.

"What year?" Daddy would ask.

"'Fifty-nine," the man would say, and Daddy would pause, thinking hard.

"B'lieve I might," he'd say, turning to my mother. "Cook, where did we leave that Ford I bought from old man Peacock? Ain't it down at the end?" "Cook" was what he called Mama, affectionately intended; her given name's Lee Ada. She might remember and might not, but they'd gather up handfuls of tools and go hunt a radiator.

For my father, the junkyard was more than a business. He was thrifty and shrewd and could spin gold out of straw, so salvaging completely suited him. He threw nothing away; he kept everything he found; whatever he was given, he took. He saw value in things most of us wouldn't glance at twice, and more importantly, he could fix things.

I've seen him haul a vacuum cleaner or a fan from the dump and have it working in an hour.

Besides, Daddy is a mechanic in the word's truest sense, loving motors and tractors and radios and guns less for their implementation than for their fabrication. He hunts the bolts, cotter keys, wires, shafts, and belts that hold together metal pieces, engineering usefulness. He is on the trail of a sprung spring or a broken part. I have seen him sit for hours with loupes strapped to his eyes, taking apart a railroad watch, lifting with tweezers gears and screws no bigger than atoms. His game is understanding and order, two things denied him early on.

Curtis Hamilton was one of Daddy's few friends, probably because he was older and earned the respect mandated elders in the Bible and because his high IQ approached Daddy's own. He'd been a state senator and, although retired, ran the cotton gin in town and was always inventing some piece or another for the gin and needing a particular piece of metal to do so. I remember Mr. Hamilton by the sunflower-like Jerusalem artichokes he grew, whose edible roots he pickled in pint jars. He was a curious man, an experimenter and an inventor, and when he and Daddy put their heads together, sparks flew.

Curtis told Daddy about a math problem that he had given many people over the years. In his entire lifetime only three people had been able to solve it, a retired old-timey teacher, a Jewish merchant, and Hamilton himself.

The problem went like this: *A man had two grandchildren, aged ten and fourteen. He had $10,000 to will the two of them and wanted each of them to inherit an equal amount on their twenty-first birthdays. The money would be deposited in two* *accounts with a fixed rate of interest, say 6 percent. How much of the $10,000 should go into each account in order for the two children to inherit equal sums?*

It is not an interest problem. It can be solved by trial and error, but the rules specify you must have a formula. No one can help you.

My father, although a high-school dropout, had pocketed the equivalent of a diploma from a business school downtown and had received his grandfather's instruction in mathematics—sine and cosine, square root, degree of angle—while learning to survey.

For five days he labored day and night for an answer. He would not rest until the problem was solved. After the first day, he telephoned Mr. Hamilton.

"You're on the wrong track," Mr. Hamilton said. "Wrong track."

On the fifth day my father telephoned again to present his answer.

"Did you have any help?" Mr. Hamilton queried.

Daddy said, "Not a bit."

"Have you been working all this time?"

"Off and on."

"Well, you got it," he said and reluctantly inducted Daddy into his club. Later Dell gave the problem to our high-school algebra and precalculus teacher who had the education to solve it in about ten minutes. Daddy had to work from scratch.

Daddy tried to teach us children to think and to value knowledge, mostly by modeling. He wanted us to live a life of the mind, as he did. For years, for example, he tried to understand Einstein's theory of relativity. In the

evenings he read encyclopedias. He expected us to study
and make good grades.

Daily he asked quiz questions and rewarded correct
answers with nickels and dimes, sometimes quarters.
Many of them were historical, some scientific, some
mathematical.

"I've got a question for somebody," he'd say at the dinner table. "It's worth a nickel. Who can tell me the name
of Robert E. Lee's horse?" And usually one of us could answer, and Daddy dutifully fished a coin from his pocket.

"Which direction is up?" "Who invented mass production?" "What does e = mc square mean?" (That one was
worth a quarter.) "Who was in the duel with Alexander
Hamilton?" "Where was Eli Whitney when he invented
the cotton gin?"

Sometimes we know what a thing means before we have
the vocabulary for it and bewilderment was my precocity.
I was six the year mental illness stole my father. I was in
first grade in Mrs. Purdom's room, my teeth falling out
and pushing in, and I copied laboriously the alphabet my
sister had already taught me. During recess a scrubby boy
from the orphans' home chased me around and around
the monkey bars and swing sets. I spent most of playtime
cleaving to the door of the classroom, afraid to go inside,
more afraid to be out. I didn't know what the boy wanted
with me or what would happen if I were caught.

One day I came running off the school bus to find
Daddy praying at his long bedroom window, the one that
looked out into the junkyard. The curtains were pulled
back, and the sun shone at a slant onto his dark head. He

was begging God loudly, looking out the window as if God hovered in midair on the other side of the pane, oblivious to my footfall. I stood watching a long time then rushed to find Mama. Something was wrong with him.

She was hanging up the phone. "He's sick," was all she could say. "He'll be okay."

When I went back to the bedroom, Daddy was standing, saying, "I'm coming, Lord" to a person I couldn't see, and then he stepped through the window, cracking the pane like thin ice on a pond, snapping strips of white-painted frame. Shards of glass flew against the pine floor, tinkling, and left a barbed hole through which my father had fallen. I saw him pick himself out of Mama's spider lilies and stumble off into the junkyard, face upraised to the sun, arms high and beseeching.

Mama caught up with him deep among the junked cars and led him home. I expected her to heal him as she healed us, wrapping our feet with long strips of an old cotton sheet when we stepped on broken windshields in the junkyard, taping the bandages down, ripping them off with a quick snatch when they needed changing. I waited for her medicine to work.

I was not afraid of my father when he got sick. He was simply unavailable to me. He was in a place none of us could reach; not even all of us collectively could pull him back. His body was at home, he'd lost his mind—for the first time I knew the two of them to be separable.

I was haunted by the words *crazy, loco, insane,* and *mentally ill* because if craziness happened to my father and grandfather and cousins and uncles and aunts, it could happen to me. So much bewildered me. Through the years I had no way of understanding what was happening—no language for it. I had only emotions. Terror. I was frightened by an unbidden and uncontrollable illness that might pounce at any time and leap away with me in its claws. If I could not fall asleep, I began to worry. I knew there was no way to talk sense into people who were sick—I had seen that attempted many times—so how then were we to reach them, especially someone as vital as my father. If I took sick, how would they reach me? Perhaps we had not been able to bring my father back because he was so much bigger and stronger than we were, but if it happened to me, surely he could wade in, throw me across his shoulder as he had done when I jumped off the roof and sprained my ankle, and drag me from the lion's den.

When I was grown, I asked my father about his experience. He replied in a letter.

> *Mental illness, or nervous breakdown as some call it, is nothing to be afraid of, or to put it in better perspective, nothing to live in fear of. In some ways it is like death. Natural death comes in stages—stage one gets you ready for stage two and so on until the coma. Most of the time people with a mental disorder don't know it. Sometimes their friends can't detect it. Close family members usually can, as well as those trained in the field.*
>
> *Thirty years ago I had what people call mental illness. I call it one of the greatest experiences of my life. I would not erase it from my past even if I could. I would not sell it for a million dollars. Its value to me cannot be measured. I can only assume that God allowed it to happen and was with me all the way through it—one in the Church said mental illness is of the devil, which I do not agree with.*

It taught me: 1) greater love for people. 2) greater love for the earth, the trees, the hills, the valley, the streams, the soil, the animals. 3) the future is everything. 4) my wife is me. 5) to love my family. 6) the true value of my sanity, my health, my well-being. 7) to respect our Creator. I will not list the minuses because everybody knows what it would be like to be called crazy.

I have nothing but praise for the state of Georgia's mental institution. From what I saw, top to bottom, it was good. In closing, I would like to remind you of what our Creator said many times. Fear not.

Crackers

Although the Southern backcountry was lawless and dangerous in the early eighteen hundreds, families poured in, displacing the Lower Creek Indians who hunted the pine country between the Altamaha River and Florida. Most of them were Borderlanders, coming from the Celtic-influenced regions of Britain, predominantly the borders between England and Scotland, but also from the Scottish Highlands, Wales, and Ireland.

Peril was an old companion for the Borderlanders. For seven centuries, the kings of Scotland and England had disagreed about the ownership of the Borderland, and the region was constantly in upheaval, under attack, or at war. Because of their tumultuous and violent history, Borderlanders tended to be migratory. They were ready for bloodshed. They banded together for defense into familial clans spanning four generations, and in their clannishness, they were antipathetic toward strangers and outsiders.

In the southern United States was room enough for clans to continue their farming and herding economy, far from government interference, where they could rule

themselves and worship as they wished. They could, as
someone said, "kick up their heels and wear out their
knees." Because of the area's remoteness, their culture
did not dilute, and they were able to retain many of their
customary ways.

"Such a region was ideally suited for the clannish, herding, leisure-loving Celts," noted Grady McWhiney in his
book *Cracker Culture: Celtic Ways in the Old South,* "who rel-
ished whiskey, gambling, and combat, and who despised
hard work, anything English, most government, fences,
and any other restraints upon them or their free-ranging
livestock."

History had rigged them to be Southerners.

The people were called Crackers. A possible derivation of
the term comes from its meaning *boaster, braggart; hence,
a liar,* as when Shakespeare writes, "What cracker is this
same that deafes our eares with this abundance of superfluous breath?" One G. Cochrane, in a letter of June 27,
1766, wrote, "I should explain to your Lordship what is
meant by crackers; a name they have got from being great
boasters; they are a lawless set of rascalls on the frontiers
of Virginia, Maryland, the Carolinas and Georgia, who
often change their places of abode." Others claim the
name came from the settlers' cracking of whips over teams
of oxen or mules or even for their habit of eating their
corn cracked. The term began to refer to poor Southern
whites.

Cracker speech, called Southern highland or Southern
midland by scholars, developed from that spoken in the
British Borderlands: patterns of pronunciation (*whar* for
where, *pizen* for poison, *shet* for shut, *young-uns* for young

ones, *winder* for window, *Toosdy* for Tuesday); a distinctive
vocabulary (*fixin* for getting ready to, *scoot* for slide, *swan*
for swear, *haint* for ghost, *rinctum* for fuss, and *honey* as a
term of endearment); odd verb constructions *(they growed
up);* and widespread use of the double negative.

In January of 1818 on the Flint River, the Creek Indians,
having been bloodily defeated by Andrew Jackson in
their 1813–14 rebellion against the encroachment of
whites, ceded a tract of land below the Altamaha to
Georgia. Until then, the river had been the dividing line
between European and native settlement, one whites
dared not cross. The Georgia legislature lost no time in
taking over its geopolitical spoils, and in 1818 created
three new counties, one of which was Appling. The land
below the Altamaha was open.

The Crackers entered a region dense with longleaf,
sometimes called yellow pine, Georgia pine, or most
commonly, heart pine. Because wiregrass covered the
forest floor, the area came to be known as Wiregrass
Georgia, and its settlers were said to be as tough and wiry
as the grass. They felled one hundred-foot trees to con-
struct crude heart-pine cabins in a typical style, called
shotgun, which meant a slug could pass straight through
the cabin—in the front door, through a hallway that bi-
sected the house, and out the back. This was known as a
breezeway. Sometimes two doors opened out the front
porch and a separate kitchen connected to the living
quarters by a walkway called a dogtrot.

Cracker settlers built split-rail fences, barns, smoke-
houses, and chicken pens. They kept the yards around
their shotgun cabins raked to bare dirt, to keep them

from torching, and they cleared holes in the wilderness
for fields. They cut timber from their lands and rafted it
down the Altamaha River to Darien.

In their new home, oats would not grow, but corn
flourished. Oat porridge turned into grits, boiled
hominy made from ground corn. Swine and cattle re-
placed sheep. The meaningful war games of the past
became the sports of wrassling (fighting) and target
shooting.

Some squatted; otherwise, land was distributed to vet-
erans of the Revolutionary War in the lotteries of 1820,
1821, 1827, and 1832, and by 1850 all the Lower Creeks
had been dislodged from their domain and sent to reser-
vations in Arkansas.

My mother, Lee Ada, is the daughter of Beulah Miller
and Arthur Bowen Branch, who was the son of Mary
Parker and Walton Branch, the son of Susannah Baxley
and Michael Branch. Susannah was the sixth child of
Mary Johnson and Wilson Baxley, for whom our town
is named. Wilson begat Susannah begat Walton begat
Arthur begat Lee Ada begat me.

Daniel Johnson was Mary Johnson Baxley's father and
my great-great-great-great-grandfather. He is listed as a
resident of Appling in the 1820 census, taken two years
after Native Americans ceded the land to Georgia, and
the county was created.

We have been here for a hundred and eighty years.

In 1826, the year he married Mary Johnson, Wilson Baxley
was a cattleman. His herd ranged through the unfenced
pineywoods, and every spring those cows ready for market

were rounded up and driven to market in Savannah. When the roundup crossed the river, boats flanking the herd kept the cattle in the crossing.

Wilson Baxley was so strong he might break your jaw if he slapped you. Once at the Altamaha River he met two brothers who had heard of his strength. One of them said, "I bet my brother can take you down." Baxley immediately threw the man and then, because he could, stripped off his clothes and swam across the river, a distance of five hundred feet. It is said if he had business at the state capitol in Milledgeville, he walked the entire distance.

The lives of the settlers were not without hazard, self-inflicted and otherwise. In one handed-down story, Michael Branch, Wilson's son-in-law and my great-great-grandfather, had gathered a bunch of men at the river to help at cow-crossing time. One good swimmer had to get back from across the river and did not have a boat. He plunged into the current and was making progress when supposedly an alligator started in pursuit, or appeared to. Hollering, the men warned him and he began to swim as fast as he could, heading directly for the far shore. It was a tight race. Luckily for the man, the men on shore started shouting and splashing in the water, distracting the attention of the alligator and slowing him enough for the man to win the race.

Most important about the longleaf pine settlers was their attitude toward nature, although this can only be inferred. They were great hunters and fishers, great woodspeople. When they entered Appling County, it was part of a rich wilderness stocked with fish and game, and it lay at their disposal. They had no thoughts of a future—in America, as there had not been in Britain, a great frontier lay beyond. They hunted; they fished; they slew natural predators. They shot passenger pigeons by the hundreds out of the skies.

"In no place during the antebellum period," wrote McWhiney, "was it easier for Celts to maintain their traditional pastoral ways than in these great forests that covered much of the Old South." Into this new country, full of roots and berries, they turned loose their branded stock to forage as best it could, in an open-range method of herding they brought too from their premodern Celtic culture. This earned their hogs the name "piney-woods rooters." If they gave any thought to the environmental destruction wreaked by the introduction of armies of foreign beasts, it went unrecorded, dying as all conversations—over split-rail fences, on the drover's trail, at revival, at hoedown—died, forgotten or never said.

In 1858, a traveler from Brunswick to Eastman, identified only as R. J. M. in his letter to the *Brunswick Advertiser and Appeal,* described Wiregrass Georgia:

> . . . only at long and tedious intervals did rude pine log huts present themselves with the most forbidding surroundings—naught lovely, nothing inviting in all this vast solitude, save Nature clothed in the majesty of her glory. On every side, far beyond the range of human vision, was a continued pine land carpeted with green spread out in panoramic beauty. Grove after grove of majestic trees of fleecy foliage, with now and then a pine larger and taller than the rest, standing like a grand old patriarch of the forest, who had for ages in silence stood with uninterrupted vision guarded this wide expanse of younger growth. These grand old

trees with their hundreds of concentric circles of wooded tissue, which had been forming quietly, regularly and mysteriously for ages, were high enough and stout enough to make masts for the largest vessels, or lumber sufficient for a common-sized house. Through this whole country no roads penetrated, save what had formerly been deer-trails, and subsequently cow-paths.

By the time the traveler returned in 1885, the place was unrecognizable. Within a quarter of a century the grand woodlands were gone.

"Gone!" he wrote. "An invasion of a terrible army of axemen, like so many huge locusts, has swept over the whole face of the land, leaving nought of former grandeur but treeless stumps to mark the track of their tramp."

My kin lumbered across the landscape like tortoises. Like raccoons we fought and with equal fervor we frolicked. Because we needed room, our towns sat far apart, often thirty miles. Accustomed to poverty, we made use of assets at hand, and we did not think much of prosperity. Like our lives, our speech was slow. We remained a people apart.

More than anything else, what happened to the longleaf country speaks for us. These are my people; our legacy is ruination.

Native Genius

Daddy's was an amazing triad of traits—frugality, creativity, and mechanical ingenuity—so that as I grew, our estate grew. Junk bred junk.

I know now my father's occupation has an actual title; he is a *bricoleur,* a term given by French anthropologist Claude Levi-Strauss to folk recyclers, people of creativity, vision, and skill who use castaways for purposes other than those originally intended, sometimes for art. Theirs is a native genius—as Joe Graham explains in his paper about *milusos,* meaning thousand uses, of Mexico—that goes beyond simply making do with what one has. Native geniuses are "able to take the materials and technology at hand and solve complex problems."

"Let me show you how I get corn off the cob," my father says, inviting me outside. He shows me a sharpened length of metal pipe six inches long, through which he slides the plump ears, their kernels falling into a pail. I'm impressed.

"That beats cutting with a knife, looks like," I say.

"You bet."

Daddy made an arbor for the grapevine behind the shop by welding together truck driveshafts, hollow metal rods that connect engines to tires, like iron logs. He formed a skeletal box out of them, with the grapevine growing inside center, and strung fence wire across the driveshafts for tendrils to catch on, like a net ceiling. From beneath, the arbor was a secret and green room, ranks of leaves so thick only fragments of sunlight passed between their jagged margins, cliffs of vines like walls hanging almost to the ground.

We could stand underneath and pick dangling grapes, but our favorite thing to do was hoist off an old car parked nearby, chin up on a driveshaft, and crawl into the leafy joy of wire and vine. There was no falling through. From on top, the arbor was like heaven, high and surrounded by all the ripe grapes you could eat. The grapes were scuppernongs—a domesticated kind of muscadine whose Indian name means "place of the magnolia"—and they'd be all around us, champagne balls about the size of radio knobs hidden among a mass of green vines and leaves. Eating them, sweet rushed against the mouth over and over. Mostly we swallowed juice and pulp, seeds and all, and threw the tough, empty rind at the nearest sibling.

Underneath the grapevine the ground was scattered with old leather boots and shoes, warped and twisted with rain and hard sun. That was Grandpa Charlie's idea. Grapevines need a slow fertilizer because they quit bearing if heavily nitrogenized. Shoes, being leather, were rich in nitrogen and since they decomposed gradually, according to his theory, the fertilizer would slowly enter the ground beneath the arbor, enriching the vine without overwhelming it. It looked like a shoe dump.

When he built guns, Daddy manufactured firing pins out of old Chevrolet push rods and flat gun springs from Ford door-handle springs. The swing set we played on he welded of pipe. He cut glass with nothing more than a table or other flat surface, a bottle of alcohol, and a glass cutter. He fit a Buick piston in a John Deere tractor.

"Did it work?" I asked him.

"Couldn't sell the tractor without it," he said.

He built a pair of indestructible firedogs out of lengths of driveshafts, packed with sand, welded to connecting rods from an old engine. He used flat leaf springs for prising up tile and engineered a nut-removing tool for Uncle Percy out of a lawnmower blade and two or three pieces of flatiron. When six-lug, fifteen-inch tires were hard to come by and in high demand by farmers, needed on trailers that hauled heavy loads of corn to market, he welded the six-lug centers of common sixteen-inch tires into five-lug fifteens. He could make anything, fix anything.

"Genius, sometimes inspired by necessity and encouraged by opportunity, is not satisfied with merely the status quo," writes Graham. "What is clear is that genius, whether it is represented by a Thomas Edison or an Albert Einstein or an Alberto Ramirez, leaves the world a better place."

The day my father was taken sick he had gone to a gun show on Jacksonville Beach with an older man he had not known long, a Mr. Paschol, a white-collar regional manager of Goodyear Tires who had befriended Daddy, daily stopping at the junkyard to talk about their common passion for guns. Paschol owned a sackful of old and beautiful pistols. He was in charge of franchises in three states, but his south Georgia business was so lucrative he even took a room in Baxley. One Saturday he invited Daddy to ride to the Jacksonville Beach Gun Show with him—Daddy thought to do a little trading.

They parted in the parking arena, leaving their lunches in the car and arranging to rendezvous at the noon hour to eat. Daddy said that after lunch he began to feel unusual sensations. He felt shaky, his insides turning to gelatin, then shakier, as if he operated a noiseless and invisible jackhammer. He couldn't calm down. His heart sped up, beating like a crazed vulture inside his chest. By the time Paschol delivered him to his door, he no longer controlled much of his body, the mind chopped from it the way you'd chop a chicken's neck, leaving the carcass to go dancing off in its manic convolutions of nerve endings. He had begun to hallucinate.

The suspicion, of course, is that the episode was caused by drugs, since this occurred in the flowering of the late sixties. The other suspicion raiser is that Paschol, who had visited daily, was never seen or heard from again. Even the sheriff agreed that it was odd. Genetic predisposition was undeniable, but a hit of LSD could've been all that was needed to send Daddy over the edge of a three-year cliff.

I don't know how many weeks Mama bore Daddy's sickness, hoping he'd turn back to us as surely as he'd turned away, but in the end she had to give up. I think it was the day he locked Mama and us four children in the back bedroom, which my sister and I shared with a six-foot chest freezer. That day, he wanted us to be quiet, and we were. We sat where he told us to sit. Each of us obeyed him perfectly and immediately, watching wide-eyed as Daddy ranted in what I now know to be mania.

I have not wanted to ask my mother how long we were there, all those years ago, but I know it was hours. We began to be hungry and to beg for food, but my father would not allow anyone to leave the room. No matter how rationally my mother spoke, how gently, as if she crooned to a sick lamb, my father said *no*. If we cried, he scared us to a hush. The telephone rang and rang between the empty chairs in the living room.

It was as if Daddy knew that he was losing what was most precious to him and wanted to hold on to it. He wanted to gather up this family he loved and lock us away with him, in this mind-gone place he'd been taken. He was no longer in control, but he didn't want to be in that place without this beloved wife and these children he would gladly lay down his life for. Toward evening he permitted Mama to reach into the freezer with her eyes closed and choose one thing for us to eat.

"Honey, you know we can't eat raw food out of that freezer. Let me go cook us some supper," she protested.

"One thing," he said, "No more no less. I'll hold the lid for you. You have to close your eyes. That's the way God says to feed the children. One thing." Everything Daddy said was a rant, a poem, a list of possibilities, a monologue with the nether world. He was out of his mind.

Like my mother, I knew what was in the freezer: hamburger patties she shaped from ground beef and froze, quart bags of blanched field peas, sliced okra, a chicken or two, bags of turnips, stewed tomatoes from Grandmama Beulah's garden. Raw and solid cold.

In summer, when peaches are plentiful in Georgia, my mother puts them away for winter. She'll spend all day peeling a bushel, cutting the coral-colored flesh into slices, tossing aside the pits. When she fills a dishpan, she stops and transfers the slices to freezer bags, upright on the counter. She mixes sugar water and pours it over the fresh fruit, making peach-colored bladders, twist-tied. Then she carries them gently to the freezer.

When the lid was raised a few inches the day Daddy locked us in the back room, my mother stuck her arm inside. I can still feel on my face the cold vapor that billowed from the white crack. Our eyes watched her as she withdrew a package, brushed away a skift of frost. It was peaches.

Looking back, I wonder if Mama had remembered where she'd piled the fruit or if her blind fingers had roamed desperately among the icy piles, feeling the awful sameness of the bags. Was it utter luck? Later I tried to do what she did, sticking my arm into the freezer, the heavy lid pinching my shoulder, feeling in the burning cold until my fingers stung. When I'd open my eyes, I'd be holding chicken necks or turnip roots.

My grandmother, knowing something was awfully wrong, rescued us by calling the sheriff who sent deputies for Daddy. Kay says she remembers them putting a white sheet over his body, even over his face. He stayed overnight at the county jail, and the next day was transported to the state hospital at Milledgeville. Mama was at the jail when he left.

She told us that he cried when they left with him. The last thing he said was, "Take care of the children."

On the way to Milledgeville, Daddy slipped into a coma from which he did not wake for four days. When he came to, he sat up on a bed in a white room, rubbed color back into his face, and combed his fingers through his wiregrass hair. He was weak, his stomach concave. He lifted weakly to his feet and went slowly to the doorway. A man was sweeping the long corridor. The gentleman paused his push broom.

"Sir," Daddy asked, "Can you tell me where I am?"

"I sure can," the man replied. "You in Milledgeville."

"Milledgeville?"

"That's right. The very same."

"Does my wife know I'm here?"

"I don't know about that, man."

An orderly came bustling down the hall in a white coat.

"You awake," he declared unnecessarily.

"I guess so," Daddy replied.

"Good," the man said. "We was wondering when you'd wake."

"How long have I been sleeping?"

"Four days now. We didn't know if you'd ever wake up."

"What day is it?" Daddy asked.

"This is Wednesday."

By then, the bottom had dropped out of the world. Daddy did not wake up healed but stayed gone. His absence was a steep-sided quarry, filled with marbling water, and there was no climbing out of it.

Mama waited, her husband three hours and a thousand light-years away. She found the money to visit Daddy, packing us children into the car along with enough food for the day, hoping the Mustang would not break down and leave us completely at the mercy of strangers. At the state hospital the doctors told her that her husband might never be well, might never be released, and not to expect him ever to return to the family.

What faith she had. She was so strong a ship could have been hewn from her body. If it weren't for us children and her powerful mothering instincts, she would have broken, I think. Instead, she kept a vigil of prayer — praying as every pan of biscuits rose in the hot hot oven, praying as she mended a sweater, praying that the chair at the head of the table would again be filled with the man she loved. Every sweep of the worn broom was a prayer. During our visits, Daddy was his old self for minutes at a time and, as the weeks passed, these minutes expanded until they came together, like warm currents of water in the river, until all of it is summer. And then he was able to come home.

Although my grandfather took to wilderness for solace to ease his wracked mind, my father turned to machines, and somewhere, between the two of them, the thread of nature was lost. Fierceness took different forms in them, one savage, one inventive. What was balm for one was terror for the other. Not long ago I asked Daddy why he never fishes or hunts.

"Didn't Grandpa ever take you with him?" I queried.

"One time," he said, "and that was enough for me never to want to go again."

"How old were you?" I asked.

"Oh, four, five, six." He lifted his hand and held it for a moment in midair, dropped it.

"What happened?"

"He took us coon hunting one night," my father told me. "We followed him for hours in the dark. We got tired and hungry and wanted to go home. He made out like we were lost and something might attack and kill us, all such as that. He had us all crying."

"He meant to do it?" I asked.

"Oh, he got a kick out of it," Daddy said. "He liked to play tricks."

And that was that. So much for tradition. So much for a long line of outdoorspeople. So much for the woods. What my grandfather planted in my father was a crazy fear and mistrust of being lost in a wilderness alone. If there ever was a wilderness misunderstood, insanity is it.

I think of my own life, how it embraces a great quest to know every cog of nature—the names of oaks and ferns, the secret lives of birds, the taste of venison and Ogeechee lime, wax myrtle's smell and rattlesnake's, the contour of bobcat tracks, the number of barred owl cackles, the feel of Okefenokee Swamp water on my skin under a blistering sun.

I search for vital knowledge of the land that my father could not teach me, as he was not taught, and guidance to know and honor it, as he was not guided, as if this will shield me from the errancies of the mind, or bring me back from that dark territory should I happen to wander there. I search as if there were a peace to be found.

Timber

After the Civil War, lumber was in great demand for re-construction, and the war-torn country turned to the pine belt of the Southern coastal plains. About the same time, the production of naval stores in North Carolina began to wane and big turpentine producers in North Carolina sashayed into Georgia. Sawmills as well as turpentine stills sprang up, and by the 1880s the great long-leaf forests were in trouble.

As early as 1870, wood-burning locomotives seethed between middle Georgia and the coast. Railroads were to pines what they were to buffalo: the means to extinc-tion. The *Brunswick Advertiser and Appeal* reported that in 1875 the first trainload of turpentine was received at the Brunswick station, where a crowd had gathered to watch its arrival. In 1882, the same paper announced that fif-teen mills dotted the Brunswick and Albany Railroad, "with a capacity of about 300,000 feet, sending almost their entire production through this port," and that loads of turpentine arrived daily.

This paean to commerce was not without warning: "While the supply of timber is far from inexhaustible there is still a large area of yellow pine forest through the pine belt of Georgia, that is almost untouched by the ax of the timber cutter and will supply the demand of the lumber manufacturers for years. . . . At present the term 'timber butcherers' can with propriety be applied to many of the so-called manufacturers who send 50 percent of the logs to the slab pit," the editor wrote. A slab pit is the crater where the remnants, or rounded edges, of squared-up logs were burned.

A year later, a writer in the *New York Times* predicted that "the people of the Southern States will soon find it expedient to protect their forests" against lumberers who had destroyed the forests of Michigan, Wisconsin, and Minnesota. "Those who are at work in these Southern forests, like those who have almost exhausted the forests of the lake region, care only for the lumber and the money that is paid for it. The inevitable effect of their work upon the climate, the streams, and the soil is something in which they take no interest, but it is a matter of great importance to the permanent inhabitants of these regions."

The traveler, identified by his initials R. J. M., who had passed through the region in May of 1858, going from Brunswick to Eastman, retraced his path twenty-seven years later, in September of 1885, and was so shocked by the changes incurred during that time that he immediately dispatched a letter to the Brunswick newspaper:

> The time we write of was over a quarter of a century ago—a long time, is it not? The greater portion of an ordinary lifetime of an active man. In this region of the country in this space of time man has been active, and the changes wrought through his agency have

been great. Now, September 1885, what is the change? Those grand forests—where are they? The trees, the grass, the cabins—where are they? Gone!

Naturalist Herbert L. Stoddard traveled from Chicago to Florida in 1893, as a young boy. In his *Memoir of a Naturalist,* he remembers the "vast stretches of cutover pinelands of southeastern Georgia—a shambles of fire-blackened and broken-off stumps as far as the eye could see. There were frequent stops beside long piles of corded pinewood to replenish the tinder for the wood-burning locomotive."

Writer, explorer, and naturalist Roland Harper reported in the *Savannah Morning News* in 1911 that one hundred thousand new settlers had entered the wire-grass region between 1890 and 1900. The population of some Georgia counties increased 75 percent during that time. Where there had not been a railroad junction a quarter of a century before, by the turn of the century hundreds of miles of railway crisscrossed the state.

It was in that period of boom that the people on my Daddy's side came to Georgia. My great-grandfather Walter Lynn Woodard, called Pun, who died seven years before I was born, worked as a lumber checker in the 1900s at a big sawmill in Lumber City. In his job he eyed thousands of cut longleafs and sent them to the blades.

In the 1920s, after the rush to gorge on longleaf had abated, Pun migrated to Appling County, where he served as tax receiver and county surveyor, both elected offices. He had a knack for figures and ran landlines all over the county. In the early '30s he passed the boards in Atlanta and became a state surveyor. He traveled all over south Georgia marking up land. Dividing it; dividing it again.

Pun was famous in Appling County both as a mathematical genius and as a sot. He threw drunks that lasted for days at a time, but on sober evenings he sat by the fire or in his chair on the porch, scrap of paper in hand, figuring aimlessly, the way others read or embroider to while the time before sleep.

Pun was married to Mattie Victoria Crosby, known as "Little Granny," a tiny, sharp woman who surrounded her unpainted, heart-pine house with flowers.

"House looked like a Russian funeral," my father said. "In those days," he continued, "they knew more about the names of plants and birds and so forth. You didn't harm a bird around Mattie. Couldn't shoot a mockingbird. Probably she was kinder to birds than people."

When my father was hungry, he couldn't depend on Grandmother Mattie for even a biscuit.

"She was very stingy," Daddy said to me. "But remember . . . she was a full-grown woman who suffered during Hoover Days."

I want to think that my great-grandfather Pun felt some love for land and trees—after all, they had been his livelihood and his life and surely his wife lobbied him to save land for wildflowers and birds. My father says no. While a young man, he had worked for Pun. "Grandpa walked the woods a lot," Daddy told me, "but not for enjoyment. Much of the land he surveyed in preparation for logging."

Pun used to recite a rhyming triplet to his grandson that my father was fond of repeating.

There's just as many fish swimming in the ocean today
luscious and beautiful in every way
than have ever sputtered and spewed in the saucepans
     of yesterday.

"The moral of the story, Son," Pun would say, "is *Don't take more on your heart than you can shake off on your heels.*"

Of all lessons, that one I never learned and hope I never do. My heart daily grows new foliage, always adding people, picking up new heartaches like a wool coat collects cockleburs and beggar's-lice seeds. It gets fuller and fuller until I walk slow as a sloth, carrying all the pain Pun and Frank and so many others tried to walk from. Especially the pain of the lost forest. Sometimes there is no leaving, no looking westward for another promised land. We have to nail our shoes to the kitchen floor and unload the burden of our heart. We have to set to the task of repairing the damage done by and to us.

In 1935, Pun surveyed a piece of upland on the Old Surrency Road that my parents now own and have logged and replanted with "improved slash," the fast-growing, off-site species of pine that is replacing longleaf throughout the South because of its greater commercial value. One day that timberland may be mine. It will be truly and unforgivably mine.

Heaven on Earth

When I was young, religion was the rock foundation on which our lives were solidly constructed. More than a life of the mind, my father desired a life of spirit. Daddy had started his own church when we were small, but abandoned it after a year, since God had not officially and personally called him to preach. He was fundamentalist, fervent, holy-rolling, hungry to explain human existence on the planet, and eager for a reason to live. God had put us here and given us the Bible as a field guide, and my father would serve him.

I wasn't allowed to wear pants or cut my hair, wear jewelry or makeup. When I prayed, my head had to be covered. My family didn't participate in worldly activities—didn't celebrate Christmas or Halloween or Easter, for these were pagan and of the world. We were of God. We didn't go to ball games or parties; friends didn't come home with us, and we didn't go home with friends.

We had no television. My father, not wanting the knowledge of violence accessible, had thrown it out when my sister was a baby. It was an admirable thing to do in 1958

and anytime, for that matter. What fraction of what percent did we belong to, TV-free kids?

Daddy wouldn't subscribe to a newspaper, either,
and hardly allowed one in the house. Once he burned a
discarded copy of Manson's *Helter Skelter* the library gave
us, and another time punished my sister for reading
Harlequin Romances. I read them, too, but secretly, when
no one was home, hiding them in the closet. We would
come home from the library walking funny, bulging in
odd places, holding our stacks of books nonchalantly to
our chests, questionable volumes sandwiched between
acceptable ones. We streaked to our rooms to unload.

We couldn't swim because we could not show our
bodies; couldn't compete in sports, couldn't date. We
couldn't cuss, couldn't dance, couldn't drink, couldn't
smoke—we couldn't even pick up a cigarette butt and *pretend* to smoke. Couldn't bet, couldn't gamble, couldn't
show our arms above the elbows or our legs above the
knees. Nudity was a big deal. We were careful about keeping doors closed while changing, knocking on bedrooms
before entering, keeping covered. The only naked bodies I saw were African and South American tribespeople
in *National Geographic* magazines. We couldn't even wear
sandals.

Every meal began with grace, which we called the blessing, and a trip around the table in order for each of us to
recite a Bible verse. I thought it cheating to repeat the
same ones over and over, if the purpose was to learn
verses, but we did. The shortest verse in the Bible has two
words and it was well used. "Jesus wept." Why would Jesus,
the son of God, be weeping?

"The Lord is my shepherd, I shall not want," also led

us into many a meal. The trouble was, there were six of us, two adults and four children, and the verses started with Daddy. If you were sitting at his left hand, the shorter verses got used before your turn rolled around, and you were stuck trying to remember one of the beatitudes, "Blessed are the pure in heart: for they. . . shall see God? Shall inherit the earth?" or grabbing a pocket Bible the Gideons passed out every year in front of the grocery store and hurriedly finding a verse that made sense and stood on its own, like "In the beginning was the Word, and the Word was with God, and the Word was God." or "Let every thing that hath breath praise the Lord. Praise ye the Lord."

If Grandpa ate with us, he sat very seriously through the blessing and through Bible verses until it came his turn.

"Jesus wept, Moses slept, and Peter went a'fishin'," he'd say, then look straight at one of us and grin, his green eyes twinkling like soda water. He did it to goad Daddy. We put hands to mouths to stifle giggles, burbling out the last of the verses. After a while we came to expect and appreciate Grandpa's sacrilege.

Late one night, listening to FM radio, Daddy heard a Philadelphia preacher whose words gripped him: "Let us have the truth, the whole truth, and nothing but the truth." The man was Bishop Johnson, believed by his followers, who called themselves Apostolics, to be the thirteenth apostle of God. Soon after that broadcast my father boarded a train in Jesup and rode to the church headquarters in Philadelphia where he was baptized and, once home, began to shuttle his young family to Brunswick two hours away, where there was an Apostolic church.

Brunswick was a dirty port town back then that smelled like rotten eggs because of the paper mill. The whole town stank. Nothing in it was higher than two stories and everything ill kempt. I never saw Brunswick lively but only on Sunday, closed-down and grim. We drove into the poor, black section of town, where children played in the street and idle men congregated in knots at corners watching women in hot pants strolling in and out of the Jiffy Marts. In a line of closed-up, abandoned businesses stood the church, painted white outside, with its name stenciled in blue on the half-facade: Church of the Lord Jesus Christ of the Apostolic Faith. It once had been a shop of some sort—shoes, groceries—which made it affordable for the meager congregation and allowed passersby to look in the shop windows. The view was blocked by a wooden partition and the simple nave was painted white, the concrete floor a dark green.

Even in worship we were different, not only because the Apostolic following was small and in retrospect marginalized, but because it was a black church. The apostle was black, the deacons, the congregation. We were white. We had been taught that the color of a person's skin was not a measure of his or her heart, but in church we stuck out like sore thumbs. The doctrine forbade wearing bright colors, only navy blues and browns and blacks, which accented the chalkiness of our skin. I say skin, but really only our faces and hands were showing. Once at a fellowship meeting in Macon, where the congregation was much larger and a magnificent choir echoed the songs of wood thrushes, someone took a snapshot of us: six white faces in a sea of black.

Often there would be less than a dozen people in the

Brunswick congregation, which allowed no leeway for
squirming or playing. Brother Randolph led the service,
and it was long and included a taped sermon from Bishop
Johnson or, after he died, Bishop Shelton. Brother and
Sister Randolph had four children too, about our ages,
and we made eyes and grinned at each other, although
none of us dared move or make noises.

The drive to Brunswick was an hour and a half in a
car crowded with four children attired in dark polyester
dresses and drab little-boys suits, and by the time we got
to church, we would be hot and tired and irritable. The
service was long. Sometimes I fell asleep before it ended;
if anyone noticed, they waked me.

Church is a glorious place to daydream because no
one knocks you from your reverie with a question to answer and no one asks afterward what you learned, which
is good because you understand hell is burning but further expostulating—the Father, Son, and Holy Ghost
being one entity, for instance—baffles you. What is entity? Imagination, on the other hand, keeps you awake
and occupies your time.

In church I had this one recurring daydream.

All my life I have loved babies, and my maternal in-
stinct as a young girl was even more powerful than when
I had matured. I was always toting dolls around in my
arms, on my hips, across my bony shoulders. I had one
baby that looked completely real. She was just the size of
a newborn, with skin made of soft latex instead of the
usual hard plastic and eyelids that fell shut when she was
laid prone. Mama saved baby clothes and diapers and
blankets from her four children, and she loaned them
to me to dress my babies in. More than once people

thought my baby was real. Once I came into the living room with my doll swaddled, head barely showing over my shoulder, having heard a man's voice I didn't know. A customer waited while my mother wrote a bill of sale for Daddy to sign.

"Mrs. Ray," the man said to my mother, "I didn't know you and Frank had a new baby." Mama laughed in her incredibly kind way and said, "No, that's my daughter's doll. It looks so real people expect it to start crying any minute." I'd be tickled, if embarrassed, and lower my baby to show him, shake her a tad to prove her inanimateness.

Kay and I had two dolls that looked the same. Twins, like the first cousins I was named after. They are a year older, named Janet and Janice. My sister called them "the two Janneices," and upon being asked when I was found what the new baby should be called, of course my sister said Janneice. All babies were Janneice. The nurse writing the birth certificate misspelled it, which led to my name being prounounced one way at home and another at school.

We called our twin dolls Roberta and Rebecca and never played with them because we wanted to keep them nice.

I daydreamed about having a real baby, a fantasy that occupied me mightily during church. By this age, nine or ten, I had determined that, despite the stories my parents told of our own births, babies grew in their mother's bellies, and my belly was just big enough to grow a Lilliputian baby, one four or five inches long, about the size of gray squirrel offspring. I could keep her in my urchin-sized red plastic pocketbook, use that as a cradle and a carrier if I took out my brush and handkerchief and left my billfold at home. It would be dark in there but babies mostly sleep anyway. If she cried she would have such a puny cry that no one except me could hear her, and I would snap open the pocketbook and give her a bottle.

The bottle would have to be small, the kind made for dolls, in fact. I would have to buy a lot of things special for her. People would think I was buying them for my dolls, never suspecting I had a real baby. I would have to steal milk out of the refrigerator for her bottles and sew her the smallest diapers in the world. Doll clothes would probably fit.

At home I could keep her in my room or I could take her outside, down into the junkyard, and rock her in the pine tree or find a junked car that was pretty clean inside to take care of her in, which is where I played with all my dolls, sitting in a dusty back seat surrounded by two or three of them, teaching them, fussing over them, holding their plastic mouths to my pea-sized breasts. They would not be jealous of the new baby, although she was for-real and they were not.

I tried to listen in church. I would get on my knees on the floor when it was time to pray, and I would take out my coin offering when the plate circulated. But there were so many problems to solve: Who would take care of my baby while I went to school? What would I name her? If she were four inches long at birth, wouldn't that mean she'd grow up a midget? People would make fun of her. People would think if I had a baby I must be married, and I was only ten years old.

I was twelve when I was baptized in the Brunswick church. Although Daddy wanted us to be old enough to choose baptism for ourselves, he was pleased when we decided it was time to give our lives to God. Mama was pleased because Daddy was pleased and because her family meant everything to her. The pool stood in one corner of the church and had a Bible verse lettered on its side: "See, *here is* water; what doth hinder me to be baptized?" Daddy passed the word to Brother Randolph, and on a Sunday night soon after, baptism service was held.

Baptism services were in the evening. Usually we attended only morning services because we had such a long drive home. My sister had been baptized maybe a year before, but I had been sleeping in my metal folding chair and missed it, and I was determined that no one would miss mine. I let Dell and Steve know ahead that no matter what, they were not to fall asleep that evening. When it came time, Sister Randolph, Sister Smith, and Mama went with me into the back to get dressed. I took off my Sunday clothes and they put a white robe around me. I came into the church, flanked by the women, and Brother Randolph led me to the pool. He waded into the water.

I climbed the steps slowly because of the robe, not daring to look out at the rows of empty folding chairs, only a dozen or two of them filled. Because it was winter and the church stayed unheated during the week to save money, the water was frigid, much colder than I expected.

I remember that it was clear and clean, however. We stood in it a long time, Brother Randolph in his black robe and me in white, trembling from nervousness and cold, as well as from the significance of the act. When I left this water I would not be the same person I was when I entered. My head was covered with a white cloth, just a handkerchief that got loaned to street women who came into church curious to see what was happening. They must cover their heads, even if with a pinned-on cloth, in the House of God.

Brother Randolph bound my wrists with one hand and prayed for a long time to God, telling him that a new lamb had come into the fold, that a young sister desired to give her life to him and live in his house, that she was ready to be filled with the Holy Spirit. He prayed that God would wash my sins away, guide me and shelter me and bless me with the gift of the Holy Ghost.

When he finally pressed his hand to my forehead and pushed me backward and down I wasn't quite ready for it. Would he forget I was under? I was freezing cold in the water, shivering in fact, and the prayer had been long. I grabbed for him on the way down to no avail, and when he lifted me, I came up blowing and swiping at my face. Sister Randolph at the organ was pumping out the same chords, over and over. Brother Randolph helped me to the steps, where the sisters were waiting with white towels.

I wanted to feel different. I wanted to start over and never sin again, not to think another sinful thought ever after that. I wanted to be the epitomy of kindness, to be long-suffering, slow to get angry at my brothers, quick to forgive. I wanted to float across the church and have the water vaporize from my robes, to grow wings and fly around the church and out the door, up into the clouds. But I would be lonely up there, and where was there glory in separation and who would mop up the trail of water behind me across the church floor?

The word we were taught for mental illness was *sickness.* "Your Uncle Nolan's sick," we would be warned, so we would not be frightened and confused by his erratic or loud or otherwise manic behavior. When he thought he was God, when he tried to walk on water, when he burned money, we would know it was the sickness. We would hear that Aunt Lee was sick or overhear my father saying, "Once when Daddy was sick. . . ." Afterward, he would refer to his own mental illness in the same terms: my *sickness*, when I was *taken ill*.

At school I learned the colloquialisms.

"Is your daddy off his rocker?" some kid would ask, repeating what she'd heard.

"We don't talk like that around here," my parents reminded us. All the crazy words were banned from our house: *berserk. nuts. not in his right mind. out of his mind. loose screw. maniac. lunatic. insane. half a bubble off. demented. loco.* We weren't allowed to call anybody crazy in any language.

"They're going to have to send me to Milledgeville," a teacher would say. "I can't find my grade book." The class laughed, hoping she'd never find it, but for me the joke hit too close to home. Many of my people had been to Milledgeville. Nervous breakdown meant something to me.

It was weeks before Daddy came home, and while he was in Milledgeville, we visited him two or three times. I remember vaguely walking the grassy, well-kept grounds, with beds of blood-red salvia and pansies purple as medals of honor. I'd never seen a yard so pretty.

On the first visit, Daddy still was not himself. He stopped to speak to an old, vacant-eyed man sitting on a bench within the high, wrought-iron fence.

"This here man was nailed to a cross," my father told

us wildly. "Show 'em, Buddy. Where you were crucified."
The man extended his hands, palms up, and pointed one
at a time to scars from puncture wounds, then turned his
hands over to exhibit identical scars on the backs.

"Show 'em your feet too." It was the same—nail wounds
through his arches. I asked the man why he'd been nailed
to a cross like Jesus and had they put thorns on his head?
But he couldn't make sense of it for me.

I was different. It was not with great distress, not until
later, that I longed to be normal, because I was so well
shielded from the world that I did not know what normal
was. I didn't know what I missed. I was well loved. Very
well loved. We were isolated from the world, but we had
each other. We were constantly reminded of our bless-
ings: health, enough food, a place to live, parents who
loved us beyond reason.

"Be thankful for what you have," my father reminded
us at every turn. "You could've been born in Haiti. Children there go to bed hungry, crying because there's
nothing atall to eat. They die of starvation." Or he'd say,
referring to his own childhood, "Think what your lives
would be like if Charlie Ray had been your father."

Yet sometimes I felt discontent, not an overwhelming
unhappiness, but an absence of lightness. I was bewildered. I would feel as if something dreadful had happened, and I carried the burden and grief of it, though
for years, until the death of my Grandpa Charlie when I
was sixteen, life spared me extreme loss. Nothing had
happened. These feelings left me disoriented.

Often with visitors the conversation turned to religion,
and when religious fanatics find each other, all else is

forgotten. My father would argue for hours about the
Bible. He knew it frontward and backward and was vehe-
ment in his belief, not to mention brilliant in his argument, and so dogmatic that I continue to associate strife
and disagreement with Christianity. If he was arguing
Scripture, meals and sleep got ignored. I dreaded the
subject my father craved.

Often people became offended, as happened with
one set of visitors who left in a rage.

"Boy, they really got mad," Steve said.

Daddy replied to him, "If God will be my God and take
me through the night and wake me in the morning, then
I don't care if the whole world gets mad with me." Moment
by moment he traded the world for God.

He explained to us that by "take me through the night"
he meant death.

"Wake me in the morning," he said, "means I want to
come out of the grave and gather around that rose and
be among those that crown him king of kings and lord of
lords, admire him and be continually in his presence.
When that time comes we can talk without speaking; we
can travel without walking; marriage, pain, time, tears,
sickness, the sun, and death shall be no more."

And we believed.

Apostolic law was more stringent than the religion my
father on his own practiced: two days a week we fasted,
which meant we didn't eat or drink until sundown. For
some reason that stricture included water. "Then shall my
disciples fast," Jesus had said in the Bible; anyone who belonged to Jesus fasted. Young children weren't expected

to abstain from food in a spiritual quest, but olders ones, except on school days, must mirror their parents.

A day without food is a long day, and I watched the clock. If you stayed occupied, time passed tolerably fast, although the breakfast hour, when the body did not receive sustenance it expected and the stomach began to shrink, was hard. During early afternoon, time crawled. Your brain ached, you got nauseated and felt very weak. By 4 P.M. the women began to cook, for on fast days supper was never late.

During summers there were forty-day fasts, honoring those of Jesus, meaning that for forty days in a row we awoke, ate nothing and drank nothing, maneuvered the entire hot day in fact with no food or drink at all, until supper. Because it was the one meal, supper was splendid, bountiful enough to make you forget the day's challenge of denial: fried chicken, biscuits, rice, gravy, fried squash, field peas, a lazy Susan of raw vegetables, sweet potato souffle, strawberry shortcake. Sometimes three of us worked two hours on making the feast. Kay started a cake early afternoon, and I joined her in the kitchen to wash and cut fruit for a salad, carefully peeling apples and sectioning oranges. During those days, no matter how hungry and weakened, I never cheated, never sneaked food or water.

Some nights the whole family gathered in the upstairs study to call Jesus's name over and over, hoping, especially our father, who craved a divine presence the most, that the Holy Ghost might appear to us. We called it a tarrying service. Everyone kneeled at a chair in the room and repeated that one name aloud. *Jesus.* In those days

I believed that we might be heard and that at any point some spirit might fill the room. Someone might be filled with the Holy Ghost as they were in church and go dancing through the house, speaking in tongues only God and his chosen understood.

The Holy Ghost was a big deal at church because you had to receive it to be saved. Brothers and sisters might ask you, in conversation, if you'd received the gift of the Holy Spirit. I would have to say no. My parents too. I didn't want it, to tell the truth. I didn't want something uncontrollable to fill me. I had seen people filled with something they said was the Holy Ghost, except it was sickness, mental illness, and I had no desire to let any such spirit steer my boat. Besides, it was embarrassing. If I got the Holy Ghost and started dancing, I knew my two brothers would bust out laughing, and we'd all be in big trouble. Or they'd get wide-eyed and never trust me afterward.

To be raised in a dogmatic, fundamentalist, isolationist religion that eschews the ways of the world was confusing for a child. I found it hard to be so different, to be the only one in my class who hadn't seen the last or indeed any episode of *Beverly Hillbillies*. To be the one who wouldn't be going to the Friday night football game, wouldn't be gathering Easter eggs, wouldn't be trying out for track or cheerleading. That the town was small made our dissidence worse: absolute.

My brothers and sister and I fantasized about the world and lied our way into it. One Christmas my brothers and I decorated a small tree in the pine grove behind the pond. Nobody ever went down there—you had to cross the pond to get to it—and the slash pines grew thick, their needles woven six inches deep on the ground, and sunlight barely penetrated the darkness beneath the lashed branches.

We had no ornaments. We made popcorn chains and draped them over the loose-branched cedar, sprinkled shelled corn on mossy stumps around. Perhaps we thought to provide Christmas for the animals, but I think it was for ourselves, although the birds and squirrels cleaned the stump of corn and stripped the popcorn chain.

On Christmas Day, we skulked to the tree with a few presents I had for my brothers, a sword carved from a stick; a gunpowder patch I'd sewn, for Dell loved to dress like Daniel Boone and lug his toy musket around; some socks filled with candy. These were wrapped in cut brown bags and the boys opened them furtively, aware that our parents weren't stupid and that Daddy might walk these woods looking for a lost sheep.

"Merry Christmas," we said awkwardly to each other.

When the holiday break was over, we made up long lists of what we'd gotten, because our school friends would ask: new socks, some books, a BB gun, a sixty-four-count box of crayons, a blouse. Sometimes we wanted these things, but more likely our lists were items we had already been given.

"In our house we celebrate Christmas year round," my father said. What he said was true. We were not lacking. He and Mama loved us with the certainty of a hurricane, and even eternity would not succeed in weakening this love. They were able, one way or another, to obtain most things we wanted and needed.

In the Book of Genesis, a pastoral male deity who

resides in the sky creates life as we know it, on the sixth day spitting into a ball of clay and forming it in his own image, breathing life into it and thus setting humans apart from the already created world he has been building the past five days. On the seventh day he rests. The earth around these brand-new humans is a stepping-stone, a testing ground, for what really matters is heaven, which lies above. In this ideology, humans are spiritual in a way that plants and animals can never be, and they hold dominion over the earth as long as they bow to God's will.

I will not endeavor to reconcile Christianity with respect for nature, because that has been attempted many times. What I want to describe is that when I was growing up, the world about me was subverted by the world of the soul, the promise of a future after death. Much of my time I spent seeking purity, meaning I desired to be good, to honor my parents and glorify God, in order to enter his kingdom one day. I prayed for forgiveness for lying, for reading *Catcher in the Rye* in the hiding spot behind my bed, and later, for kissing the office-supply errand boy in the darkroom. "Now I lay me down to sleep. I pray the Lord my soul to keep. If I should die before I wake, I pray the Lord my soul to take."

I lay awake after that prayer, too terrified to sleep, determined to be better. I resolved to measure up, to keep my knees covered by my dress, to tell the truth. What if I never saw another daybreak? What if this night the world ended in fire as it had once ended in flood? I could count all day and never get to a stopping place of numbers, and eternity was like that, longer than the grains of sand on all the beaches of the world. Did I want to burn for eternity, separated from my family who loved me?

For me, the chance to be simply a young mammal roaming the woods did not exist.

Clearcut

If you clear a forest, you'd better pray continuously.
While you're pushing a road through and rigging the
cables and moving between trees on the dozer, you'd better be talking to God. While you're cruising timber and
marking trees with a blue slash, be praying; and pray
while you're peddling the chips and logs and writing
Friday's checks and paying the diesel bill—even if it's
under your breath, a rustling at the lips. If you're man-
ning the saw head or the scissors, snipping the trees off at
the ground, going from one to another, approaching
them brusquely and laying them down, I'd say, pray extra
hard; and pray hard when you're hauling them away.

God doesn't like a clearcut. It makes his heart turn
cold, makes him wince and wonder what went wrong with
his creation, and sets him to thinking about what spoils
the child.

You'd better be pretty sure that the cut is absolutely
necessary and be at peace with it, so you can explain it to
God, for it's fairly certain he's going to question your mo-
tives, want to know if your children are hungry and your
oldest boy needs asthma medicine—whether you deserve

forgiveness or if you're being greedy and heartless. You'd better pay good attention to the saw blade and the runners and the falling trees; when a forest is falling it's easy for God to determine to spank. Quid pro quo.

Don't ever look away or daydream and don't, no matter what, plan how you will spend your tree money while you are in among toppling trees.

For a long time God didn't worry about the forests. Some trees got cut, which was bad enough, of course, and he would be sick about the cutting awhile, but his children needed houses and warmth, so he stepped in right after they had gone and got some seeds in the ground. The clear-cutting had come so fast he'd been unprepared. One minute the loggers were axmen, with their crosscut saws and oxen and rafts, and when he looked again, they were in helicopters.

When people started to replant, it was a good thing, but there was no way to re-create a forest. Not quickly. And the trees would just be cut again.

Before God knew it, his trees were being planted in rows, like corn, and harvested like corn. That was 1940, when the tree farming started, but it seems like yesterday to God.

Not longleaf. It was quirky in habit, its taproot cumbersome to deal with and slow-growing, so most of the tree farmers abandoned it. They could plant slash or loblolly and in twenty-five years be able to cut again.

Plant for the future, the signs said.

To prepare ground, they chopped, disked, rootraked, herbicided, windrowed. In wetter soils they bedded, plowing and heaping the soil into wide racks with

drainage furrows between. The land was laid bare as a
vulture's pate, and the scriveners came on their tree-
planting tractors, driving down new words to replace
the old one, *forest*.

The trees were planted close, five or six feet between,
in phalanxes. They were all the same age and size, unlike
the woodland that had been, with its old-growth and its
saplings, as well as every age in between. The old forest
had snags where woodpeckers fed and it had pine cones
eager to burst open on bare ground.

Because slash and loblolly are intolerant of fire, the
tree farmers, with Smokey as mascot, kept fire back.
Within ten years a canopy would close, and the commercial plantation was dark within, darker than you can
imagine a forest being. The limbs and needles of the
overcrowded pines drank every inch of sky. Any native
vegetation that survived land preparation did not survive
loss of light.

The diversity of the forest decreased exponentially the
more it was altered. In autumn, the flatwoods salamander no longer crossed the plantation to breed, and the
migrating redstart no longer stopped, and the pine snake
was not to be found. The gopher frog was a thirsty pool of
silence.

Pine plantations dishearten God. In them he aches for
blooming things, and he misses the sun trickling through
the tree crowns, and he pines for the crawling, spotted,
scale-backed, bushy-tailed, leaf-hopping, chattering creatures. Most of all he misses the bright-winged, singing be-
ings he cast as angels.

The wind knocking limbs together is a jeremiad.

God likes to prop himself against a tree in a forest and

study the plants and animals. They all please him. He has to drag himself through a pine plantation, looking for light on the other side, half-crazy with darkness, half-sick with regret. He refuses to go into clearcuts at all. He thought he had given his children everything their hearts would desire; what he sees puts him in a quarrelsome mood, wondering where he went wrong.

How the
Heart Opens

One essential event or presence can save a child, can
flower in her and claim her for its own. The French nov-
elist and humanist Albert Camus said, "A man's work is
nothing but this slow trek to rediscover, through the de-
tours of art, those two or three great and simple images
in whose presence his heart first opened."

For me, growing up among piles of scrap iron and
glittering landmines of broken glass that scattered ivory
scars across my body, among hordes of rubber tires that
streaked my legs black, among pokeweed and locust,
I attribute the opening of my heart to one clump of
pitcher plants that still survives on the backside of my
father's junkyard. I know it now to be the hooded species,
Sarracenia minor, that sends the red bonnets of its traps
knee-high out of soggy ground. In spring it blooms loose,
yellow, exotic tongues.

In fifth grade my 4-H project was carnivorous plants.
The only information I could find was a short entry in
the outdated set of *Encyclopedia Americana* we owned. On
a poster I sketched the innards of a pitcher plant, showing how its upright, trumpet-shaped leaves are lined with

downward-pointing hairs, how it lures insects through its lips with a sweet-smelling nectar. The insects can descend but never climb out again. I sliced open one of the *Sarracenia* stems to show the judges at the regional competition in Jesup that it was full of a ripe stew of insect parts—ant bodies, fly legs, beetle wings—but they weren't impressed.

The pitcher plant taught me to love rain, welcoming days of drizzle and sudden thundering downpours, drops trailing down its hoods and leaves, soaking the ground. In my fascination with pitcher plant, I learned to detest artificial bouquets of plastic and silk. Its carnivory taught me the sinlessness of predation and its columns of dead insects the glory of purpose no matter how small. In that plant I was looking for a *manera de ser,* a way of being—no, not for a way of being but of being able to be. I was looking for a patch of ground that supported the survival of rare, precious, and endangered biota within my own heart.

My brothers and sister and I worked hard, cleaning bricks and hauling junk, tearing down old buildings and pulling nails and stacking lumber, handing Daddy tools and feeding the sheep and cutting grass, nailing shingles and ferrying Sheetrock and measuring and sawing boards, and Daddy had neither the time nor the inclination to take us hiking or camping or fishing. Not a hard-hearted man, he could have paved the county with his empathy to the downtrodden and his compassion for hurt animals, although he wouldn't waste his breath offering congratulations to anyone enjoying health, happiness, and success. Nature wasn't ill regarded, it was superfluous. Nature got in the way.

One morning, out scrambling to get a tractor running, he stepped on a toad. The loam of the junkyard was rich and fertile, streaming with healthy earthworms, mole crickets, and warty toads camouflaged against the ground. They could be found in cool, moist places—under boards and cement blocks—where they burrowed to keep from dessicating. When you picked a toad up, it peed instantly.

This one squished under his work boot, belly exploding. I know he hated that, for when there was work to do, Daddy didn't stop for much, except maybe pause to pass a jug of ice water around or break for dinner.

He stepped to the door of the house and began to yell. "I need a knitting needle," he hollered. "Lee Ada, get me a sewing needle and thread." He was short-tempered as a hornet when he was rattled.

On the gray-painted porch floor my father sewed up the fat toad. I watched the operation from behind the living-room curtains because of my father's thunderous mood and because if I were any closer I'd vomit. Daddy stood in my little flower bed, where I planted whatever flowers I could afford with my twenty-five-cents-a-week, Saturday-go-to-town allowance, knee-deep in daylilies, poking the toad's guts back inside the crepe of its stomach like you'd do the popped seam of an upholstered chair.

Before the next day dawned, the toad was dead, stiff and dry. The tractor ran, thumping in time with my father's heart.

One afternoon a beagle, its identification number tattooed in its ear, got hit on U.S. 1 in front of the house. Grandpa had arrived in time to drag the dog off the pavement and pronounce it good as dead.

"Won't live to see the sun go down," he said.

My father has always loved life in all its manifestations and will fight for breath, his own and anything else's, until the bitter end. He votes against capital punishment, against abortion, against euthanasia; he votes to keep machines connected to a patient and perform any treatment that extends a life even by a day. He was a closet veterinarian, a would-be doctor.

The dog, panting on the shoulder of the highway, began howling in pain. Daddy rummaged in the medicine cabinet and found leftover painkiller the dentist had prescribed for Mama months before. He descended upon the dog and forced painkiller down her throat. The dog calmed and slept. Sometime in the middle of the night the dog began plaintive and godforsaken howls that proved it had not yet died, and Daddy got out of bed and administered more painkiller.

When the sun rose, the dog still lived and began to ululate anew. Again it was dosed, and this time, sensing hope, Daddy dragged the dog farther from the highway. In the afternoon of the second day, it lapped water from a bowl, looking up at us with a baleful expression though we dared not touch an animal in pain, and on the morning of the third day, it had risen and vanished. Or else it had crawled to a secret spot to die.

About a week later Steve noticed a dog following its master along the highway. He came running to the yard.

"Come look at this dog," he said excitedly. "I think it's the same one."

"What one?" I said.

"Just get off your butt and come look. The one Daddy doctored." Dell saw the excitement and bolted after us.

The man had almost passed the house but we could tell the dog was the same, a droop-eared, brown-and-white beagle. It walked slowly, head down, behind a man who also walked slowly. They headed toward town.

"That's it," Dell said, and I agreed. "It lived."

Valium came in yellow tablets, about the size of the node where a lemon attaches to a tree. Daddy kept them in the medicine cabinet above the bathroom sink, and he'd use his pocketknife to chop them into halves, then quarters, weaning himself. He despised dependency: having smoked through his teen years, he threw his last pack of cigarettes in the trash after he married Mama and said he'd never smoke again. And he hasn't. He did the same with alcohol, then television. A few times he quit sugar, whole-hog, just stopped eating dessert and drinking sodas.

He needed the valium.

For three years, from 1968 to 1971, Daddy battled mental illness the way the Allied forces fought Hitler. Against his doctor's wishes, he would lapse on taking his dosages of valium, then begin to slip away again. Mama was hot-wired to notice anything unusual in his conversations and actions and would badger him to start his medicine. Sometimes her urging worked, but twice more Daddy had to seek hospital help. What I learned during those years was to watch, too, for any indication of errancy, of losing my father, of him losing himself.

I was six, seven, eight, nine during those years, excelling at school, devouring my lessons, reading as many books in two weeks as the public librarian would let me bring home. In second grade, Mrs. Davidson read aloud

The Boxcar Children, about four orphan children who lived in an abandoned boxcar in the woods, scrounging for food. I was very sad for them. At least we had our parents.

Part of mania is the inability to sleep, and sleep became a guarded, precious thing in our house. We were never allowed to wake our father. If he was sleeping, we must be breathlessly quiet. When the phone rang, we sprang for it before the first ring stopped reverberating. If we woke in the wee hours of morning to the radio playing country music or a talk show or a Bishop Johnson sermon, our blood turned cold and we whispered meager prayers that asked for sleep to find the turn-off switch behind Daddy's eyeballs.

One time my brothers and I were playing with an older neighbor boy, Clay, in a ditch near the shop. Clay was teasing a snapping turtle with his boot, until at last it latched on to the leather.

"They mean," the boy lisped. "And they won't let go. They won't let go 'less it thunders." I looked at the clear sky and wondered how long before it'd storm. How would we get the turtle off Clay McDonald's boot?

Just then Clay lifted his foot and brought it down to earth with awful force, smashing the turtle against soggy ground. Horrified and mesmerized, we watched him stomp it to death, until fragments of shell dug out its own raw flesh, bloody and protruding. Then Clay tossed it in the ditch.

Back at the house, we told Daddy what Clay McDonald had done. His face clouded over with anger, the same face I'd seen when another boy showed my sister a pornographic magazine. Daddy had picked the boy up by his feet and shaken him, told him never to speak to his daughter again.

"Son, why did you have to kill that turtle? It wasn't messing with you."

"It was biting my foot, Mr. Frank," Clay said. "And them things won't turn loose 'til it thunders."

"If you'd left it alone, it wouldn't have bothered you," Daddy said. "You had no right to mess with it, especially not kill it." He turned to us, wanted to know why we hadn't stopped Clay. I didn't know what to say. Daddy sent Clay home and the rest of us got a whipping with his leather belt for allowing Clay to kill.

"Y'all are getting a whipping for letting him do that," Daddy said.

"We didn't know he was going to kill it," Honorable Stephens tried to explain.

"You should have stopped him from messing with it in the first place. What harm was a turtle doing? Let's go."

We followed him inside with dread, even panic-stricken, and watched while he unbuckled his belt. He never hit us in a fit of rage, but he believed if you spared the rod you spoiled the child. He didn't spare the rod. It was truly a whipping when he was done. Under threat of torturous though short-lived pain, beaten children, like slaves, dare not disobey.

"Who's going first?" Daddy asked. He snapped the leather together. I hated that sound—every nerve in the body suddenly wary and tender, fear constricting our breathing.

"I will," Steve volunteered. He was gritty. I never went first if I could help it, thinking some rage might dissipate in the preceding administration of punishment. Waiting gave me time to prepare. The further you got out of your mind the more surreal the whipping became and the less pain you suffered.

Seeing another beaten is as shameful and anguishing and lamentable as being beaten, because even a child knows that flogging diminishes the human spirit and seeks to avoid that reduction.

Steve stepped up to the couch. "Bend over," Daddy ordered and Steve bent double, hands on the black vinyl cushion, butt toward the room. Daddy looped the belt in his hand and drew back. He struck Steve across the fanny and thighs, one, two, three. Everybody in the room was counting. Four, five. We sneaked glances at Steve's profile.

His eyes were squeezed shut against the relentless and unending licks, hard because Daddy was so strong he could hang by his toes—toes, not feet—from our swing set, and because he did not temper his meting of chastisement. Steve's teeth were clenched. After four strikes, two tears expressed from his shut eyes and I knew the pain would be great.

Half of it, no doubt, was the embarrassment of being beaten, although we did not have to pull down our pants as some children were forced to do. More than anything else we steeled ourselves not to cry, the one sure way to heal from hurt, because crying was of no use in ameliorating the intensity of the whipping and perhaps even served to increase the severity.

Daddy lowered the belt. "Sit down over there," he said, forbidding Steve to crawl off like a wounded animal. The last thing you want to do is sit down. He noticed the tears.

"Straighten up," he said.

Dell went next. He made some small noises as he was being whipped but did not cry and when his turn was done he went to a chair and perched gingerly on the arm. His tail hurt.

It was my turn. The buckle rattled and I flinched, tightening the muscles in my butt and legs. Some teachers at school had paddles with holes drilled in them. I never got paddled at school.

If I closed my eyes, I associated the pain with blackness, spangled with jags and motes of rotating light, the same meteors you see if you press your hands too hard against your shut eyes. But if I kept my eyes open, trained on a black button of the sofa's upholstery, my mind remained in the day-lit room and kept the pain from entering my soul. The light from the window was a promise of cooling water filled with unknown frogs and fish and turtles that could not be coaxed from the security of algae and moss.

Prepared as I was, I was not prepared for the sting of the strap as it lashed across my skin. One. I could make it. Two. The pain was agonizing, pain on top of pain. The belt cut red welts I would later examine and nurse in the privacy of the bathroom or in the floor-length mirror on the back of the bedroom door. Another. The licks were delivered slowly and deliberately but without uncertainty. Pain shot through my body with each blow. Another lick knocked me forward. Four. Tears rushed over the dam of tight-pressed eye muscles and spilled onto my face. Once begun I could not stop them.

Clay McDonald was almost of legal age. How could I, a girl, stop him from brutalizing an animal? The taste of the turtle's blood was in my mouth or I had bitten my tongue. If I could hang on, the pain would soon be over. Somewhere in my neck a clutch of muscles spasmed. Five.

I was crying. I couldn't help it.

"They ain't no sense in killing things for pleasure," my father said. "And I expect you to make sure it never happens again." I know he meant well. He wanted us to make good decisions. Whipping us couldn't be easy for him.

He began to rethread his belt into his trousers. "Pull yourself together," he said. I turned and sat on the couch but that proved too painful and I folded my legs underneath, sitting on calves to lift and ease my backside. When Daddy left the house, back to the motor he'd torn apart, we turned away from each other and individually sought such comfort as we could muster. There was no talking. By dinner the incident was forgotten, but I vowed never to hit a child, not even my own.

The summer I was sixteen my father had a kidney stone that caused pain so inordinate it drove him to a specialist in Waycross. On the way, Mama and Daddy happened upon a live pondscoggin, which is what we call herons, stranded on the shoulder of the highway. The heron stood almost two feet tall, with a twenty-six-inch wingspan, and it was hurt, likely hit by a car. On the way back it was still there. This time they stopped. Both a leg and a wing looked broken.

Daddy drove to the nearest house. A woman answered his knock and Daddy asked to borrow four things: a potato, an old coat, tape, and a bag. The coat was to throw over the bird, the potato to jam over his bill, tape for security, and the bag to sack him up. The precautions proved unnecessary. The bird was too hurt to fight.

Daddy brought the heron home, splinted and bandaged the dull-yellow leg as if it were a flat tire, and taped the wing. He named the heron Clyde Scoggins and fed it fresh fish, grasshoppers, and minnows. Huge chunks of fish would disappear forcefully down Clyde Scoggins's long and sinuous neck, which he could swivel completely around in order to look behind him.

Clyde was a green-backed heron, a solitary wading bird that often perches in trees near woodland streams and ponds. Its back was green, darker on the crown and mixed with blue-gray lower down. Its neck was chestnut.

Daddy didn't cage the heron. He kept him in a large box at night, but the bird seemed content to perch for hours on a board laid across the box by day, so still he could have been used on a dollar bill. After two or three months of squawking in our living room, the heron's leg and wing began to mend. He would come walking into the dining room during supper.

One day we couldn't find him anywhere. After we called and hunted, he finally strode out from behind the couch with the remnants of a spiderweb dangling from his beak.

After a couple of months, Daddy built Clyde a plywood house down by the pond, a box on a metal pole (to discourage predators), and leaned a board against the pole as a handicap ramp for the bird. Daddy made daily walks down to the pond to check on the heron, which managed to climb out of his house and sort of feed in the shallows of the pond, although it was never able to refuse its ration of fish. One dawn, after observing a large water snake slithering up toward Clyde's box, my father lined the ramp with the flat, back glass from an old Ford truck, making the ramp too slippery for snakes. There Clyde lived his crippled, abbreviated life, half-captive, half-free.

His wing refused to heal. The bones themselves never fused perfectly. The feathers grew back for a time, then Clyde would injure himself and they'd fall out again. The wing was always a queer, hurt thing, awkward and earthly. Clyde survived winter, but early the next spring he died. Death is easier, less wrenching, when you know you have done what you could to impede it.

Clyde Scoggins wasn't Daddy's last avian pet. Not many years later, after days of heavy rains, he found a homing pigeon in Winn-Dixie's parking lot. It had been hit and could not fly and huddled near a spreading puddle. He caught the bird easily, noticing a metal band around its leg and its resemblance to mourning doves, and brought it home, depositing it in a box. The wing was hurt, so he splinted it to the pigeon's body and fed the bird peanuts, corn, crackers.

The pigeon ate readily but initiated no further contact. It kept quiet in its cage. Within a few weeks its wing had healed and Daddy began to strengthen it, taking it outside and flying it short distances on a jess. The bird was eager to go, yanking against the rope, but Daddy restrained it until he was certain the wing would hold.

Then one day he harnessed it and took it outside. He unhitched the leash. He and Mama were silent, watching. Daddy dashed his arm toward the sky. The pigeon expected a short flight and alighted again on Daddy's arm. It had realized its freedom but did not lift, and instead craned its head toward Daddy's hand.

"I think it wants food," Mama said.

Daddy brought its dish and fed the pigeon from his hand, offering it every food it loved and as much as it desired. When the pigeon was done, it paused.

Again Daddy flung his hand skyward. The pigeon took to wing and rose above the pecan tree, circled once to get its bearings like a blind dog in a roomful of company, then with great intent and without veering, it angled southward, every fiber of its body swelled for the flight home.

I tell these stories so that you see my father is a curious man, intrigued by the secret lives of animals, a curiosity that sprang from his desire to fix things, to repair the things of the world and make them fly and hop and operate again, and to mold his children into good people. He would with equal fury rethread a stripped bolt or solder a heat-split frying pan or patch a bicycle tire or reset a dog's broken leg or pull a tooth. I have seen him blue guns, which is to recoat the metal parts with a factory finish, build new stocks, fashion parts for them, chamber them, clean rust out of their barrels. I've seen him operate on pigs. He can rebuild carburetors and fix cracked motor heads. He helped the ewes with their lambing. He taught himself to replace stems in pocket watches and to dismantle one in order to repair a slipped spring.

Franklin surrounded himself with particulars useful to mechanical ingenuity, fragments and fractions of this and that, of everything, because any one piece might be necessary to link seventeen others together, to restore function to a broken machine. He would have agreed with ecologist Aldo Leopold, that if you are going to tinker with the earth, at least keep all the pieces.

It wasn't that Daddy didn't know or love beauty because he preferred systems. He did. He would drive us out to the farm as we called it, though really it was timberland, to walk the dirt roads to the shallow, snakey pond. Kay would pick a bouquet of flowers, and I would pull up a holly tree to replant. There was an old heart-pine house near our property that we called haunted. We never went inside it but stood many times on the dirt road looking at its lonesomeness.

A climbing, running rose grew around the front porch, the pink flowers beautiful against the sag and gap. We asked if we could pick some.

Daddy didn't say no. He said, "You know, it's a shame to pick something beautiful from delapidated surroundings. There needs to be some beauty everywhere."

Longleaf Clan

A clan of animals is bound to the community of longleaf pine. They have evolved there, filling niches in the trees, under the trees, in the grasses, in the bark, under ground. They have adapted to sand, fire, a lengthy growing season, and up to sixty inches of rain a year. Over the millenia, the lives of the animals wove together.

Yellow-breasted chat. Carolina and dusky gopher frog. Loggerhead shrike. Red-cockaded woodpecker. Brown-headed nuthatch. Blue-tailed mole skink. Striped newt. Prairie mole cricket. Pine barrens tree frog. Pine warbler. Pocket gopher.

The *eastern diamondback rattlesnake*, the largest venomous snake native to North America, borrows the gopher tortoise burrow for winter refuge. Adults average from three-and-a-half to five-and-a-half feet long, although the record length is eight feet. Diamondbacks feed on small mammals—rats, squirrels, rabbits—by lying in wait along their trails and ambushing them. Females give birth to from six to twenty-one live young in the fall.

Southern hognose snake. Arogos skipper. Carter's noctuid moth. Bachman's sparrow. Short-tailed snake.

Sherman's fox squirrel, at two feet and often topping

three pounds, is the largest tree squirrel in North
America, although it is highly terrestrial, foraging pine-
cone seeds, acorns, nuts, bulbs, and fungi off the ground.
It requires large foraging areas, as much as fifty acres.
Fox squirrels are gray-black, yellow-gray or black-orange,
but always sport a black mask, white nose, and white ears.
They prefer parklike woods for freedom of movement
and to better escape great horned owls, coyotes, and
foxes.

Sandhills clubtail dragonfly. Pine snake. Tiger salamander.
Florida mouse. Mitchell's satyr. Henslow's sparrow. Sand skink.
Bobwhite quail. Buchholz's dart moth. Gopher tortoise. Ground
dove. Indigo snake. Sandhill scarab beetle. Southeastern kestrel.
Flatwoods salamander.

As Southern forests are logged, these species of flora
and fauna, in ways as varied as their curious adaptations
to life in the southeastern plains, suffer. All face loss of
place.

Clyo

Clyo Woodard, my grandmother, married for love, but badly. After Charlie left the last time, after she paid him to leave, she would not speak his name. Keeping his eight children alive took all the rage she could muster.

What Daddy remembers from his childhood is nothing. Although the stock market crashed in 1929, rural parts of the country felt the aftershock for years, even into the 1960s in the Deep South. By 1937, the year my father was born, poor Georgians were mired in poverty, and in honor of her hope, my grandmother named her sixth child Franklin Delano. Daddy says when he was a boy there were four droplights, bare bulbs that dangled from a wire into the room, and an icebox, which was empty, in the house. He owned one set of clothes, sometimes no shoes at all. Supper, if he was lucky enough to get it, was a pot of collard greens and a cake of cornbread.

Granny collected welfare for the children for a time, until the welfare people needed Grandpa to sign some papers and managed to get ahold of him in Florida. He wouldn't sign.

"Tell her to send the children down here to me," he said. "I'll take care of them."

The welfare people thought they'd found the answer to eight hungry bellies and cut off Granny's check.

Clyo would catch a farm truck in town gathering day laborers and work all day in a cotton field. Her health wasn't good. She had lost a kidney a year or two earlier. Diabetic, her knees were bad, and she'd slide through the hot dirt of the cotton patch on her backside or drag along a small stool, so she could pick enough to feed her Nolan, Bertha, Mutt, CJ, Lee, Franklin, Johnny, and Little Gene.

She bootlegged whiskey on the side, a venture begun during Prohibition while she was still married to Grandpa. Then goods were delivered by a slender man driving a long black Ford. He backed up to the porch, got out long enough to open the trunk, hand a waiting Grandpa two jimmyjohns of shine while Granny counted bills to him, salute once, and leave. This man didn't play around. Once federal agents had come to his house looking for him. They thought to bribe his son with a quarter to tell them the pop's whereabouts.

"Give me the quarter, and I'll tell you where he is," the boy said.

"No. You tell us first, and when we get back, if you've told us right, we'll give you the quarter."

"I got to have the quarter first," the boy says. "Because when you go down there, you ain't coming back."

Grandpa struck out for the woods with the liquor. He stashed it and brought two half-gallon jars at a time up to the house. Men Granny knew and trusted would buy a half-pint from her or they would enter her kitchen, light a spoonful on fire and pay for a shot, sip it down, and rinse out their mouths with water. Granny never allowed enough coming around that her house looked like a bar.

Sheriff Wright was a friend and warned Granny when treasury agents were making a raid. Sometimes she had only minutes to hide the jars. If there was a hint of coolness in the air, they would stash the whiskey on a ledge in the chimney and build a roaring fire below. My father remembers hiding pint jars in the baby's cradle or in the outhouse while deputies drove into the yard. Another trick was to lift a fencepost out of the ground, stash the whiskey below, and replace the post.

Granny was smart. Once two men came into her yard approaching the hand pump located near the chicken house. One of them stood while the other rinsed out a pint jar at the spigot. Granny had come out to see what the men were after, and she had a funny feeling about them.

The tallest man held the jar out to Granny. "Can you fill this?" he asked.

"Help yourself," Granny said, motioning toward the pump, and walked back inside the house.

I don't know anything about the first time Granny got caught and arraigned in federal court, but the second time happened like this. The treasury agents didn't notify the sheriff they were coming. Instead they drove straight to Granny's house.

She was in the kitchen when a knock came at the door. One of the older girls, maybe Bertha, answered it. The men requested entrance, and Bertha opened the door. They entered the kitchen. Granny didn't have time to do much but luckily only a half-pint waited on the shelf.

There was a bucket of scraps on the counter and in one swift motion she unlidded and poured the half-pint in the bucket. She was still holding the jar when the men came into the kitchen. Their only evidence was a scent of moonshine, but they took a warrant out for Granny anyway. This was 1945. When Granny went to Brunswick court, the judge let her out on probation, and she never sold another drop.

She turned to cooking, ran the Greasy Spoon Cafe on West Park Avenue in Baxley. She loved feeding people well and having them brag on her cooking. She cooked hearty and rich, and served plentiful helpings; when you left her table, you were deeply satisfied. She kept milch cows and grew a big garden and enjoyed milking and planting because those things meant food for people she loved and plenty for her table.

Forever after she and Grandpa split, even when the children were grown, she still would not speak to Charlie. For over twenty-five years she wouldn't address him. He would visit their grown son, Uncle Gene, who lived across the street. Granny would see Charlie's car pull up and watch him get out and go inside Gene's house. Clyo would enter her kitchen and warm up what food she had—a slice of ham, mustard greens, yellow rice, yams, crackling cornbread—and fix a plate. One of the grandchildren might be playing in the yard and she'd call to her, "Melissa, come here."

"Take this plate to your grandpa," she'd say.

Twenty minutes later the plate would come back, via Melissa. It would be clean as a whistle. "He said thank you," the child would say. Both of them knew what the food symbolized and neither was willing to make further amends.

By the time I knew her, Granny was overweight, back and forth to doctors who prescribed handfuls of pills. She could not tell you what she was taking or what a certain medicine was for. If the doctor prescribed it, it was necessary. If one helped, two would doubly succeed. Granny's one kidney bothered her and diabetes caused her legs to swell. When she could bend down to put shoes on, her feet more often than not would not fit in them. Walking became laborious and then impossible.

Her doctors advised her to lose weight and gave her diet plans to follow. But she didn't have the willpower to lose weight—I wondered if I would inherit that, if marriage and children would wear me down until I had no stamina. Would I be fat and unhealthy when I was old?

To her dying day Clyo did the telling, not the following, of orders. She would pay to get what she wanted. The boys mowed her grass. Every couple of weeks she hired Kay to roll her hair and me to clip her toenails.

Granny's toenails grew yellow and thick and hard. First I soaked her feet in a pan of warm soapy water and dried them with a towel. I always felt like Mary washing the feet of Jesus. Only Mary dried Jesus's feet with her hair. My hair was long enough, dark and past my waist, to dry Granny's feet. I would think about that and talk to her about school while I trimmed and cleaned, reporting grades if she asked and explaining the school day. Afterward I rubbed her feet with sweet-smelling lotion. Clyo was tremendously grateful and paid me well, sometimes

as much as a dollar. What she was willing to pay for attention embarrassed me. She lived on Social Security and
supported Tritt, her live-in sister who was slow.

In the last years of her life Granny was enamored of
high-tech television evangelists. She watched them faith-
fully, believing in their powers of healing, and often sent
them donations of ten and twenty dollars. Her sons tried
to change her mind, but she was convinced that, with the
help of Billy Graham or Jerry Falwell or Jimmy Swaggart,
she might get up and be whole again.

In a last-ditch effort to save her life, my father brought
her to his house when I was almost grown. She could not
walk, but he pulled her up out of her chair, draped her
arm around his shoulders, and shuffled from one end of
the house to the other. He commanded her to hold onto
him and walk. Granny would moan and cry, begging to
be let go, to be let be, to sit down. She would beg us to
make Daddy let her be.

"I can't, I can't, I can't," she'd moan.

He put her on the doctor's diet that she had never followed. It slashed her food intake by half or more, until
she felt she was starving. She begged for food. She begged
for us to bring her glasses of tea, something to eat. She
begged.

Daddy wouldn't let us bring her anything. If she
wanted water, he would come and lift her from her chair,
and they would make a slow and agonizing trip to the
kitchen for a glass of water. Granny had smoked much
of her life and could not curb her appetite for pleasure.
Now, at the end of her days, she begged to be free of
misery.

If Clyo had no willpower, Daddy would supply her with

his own. He did not want to lose her as he had lost his father, and even as she understood that it was his bottomless love that demanded exercise and health, she cried out against him. In the end he had to let her go.

"You're killing me," she would say to him, pleading to be released, rescued. Grandpa had been dead three years—he had stopped eating, starving himself like a crushed animal, bleeding internally. Only his hands, massive against white sheets, had not shrunk.

Anyone would wonder why my father was not at his father's funeral, likewise his mother's, and would think it was some lack of love or respect. Far from that, his religion forbade attendance at funerals, an edict derived from a Bible verse that admonishes to "let the dead bury the dead." Truly, his love far surpassed his presence at a funeral.

Granny persuaded a daughter to take care of her, and after she left our house, she did not live long. My father was by his mother's side when she lay on her deathbed in 1980. She was not conscious, but she was calling a name over and over, a name my father had not heard her speak in years, in a voice sweeter than you could ever dream. She was saying it the way a lover would croon, soft and passionate. He had been her lover. Even through years of separation and denial she had loved him. *Charlie, oh Charlie, Charlie.*

She is buried beside him.

Hallowed Ground

The first time I saw a red-cockaded woodpecker was the first time I saw a real longleaf forest. I was grown. It was an April dawn in the biggest tract of virgin longleaf left anywhere, a private quail-hunting plantation embedded in the Red Hills of southwest Georgia. The light was dim beneath the pines, the wiregrass rinsed in spring dew. There were no mid-story shrubs, just acres of widely spaced pines of all sizes scattered across the landscape like children on a soccer field. Most of the trees were of a diameter of a size twenty dress, circumscribed by a rug of wiregrass, on and on.

Birds sang: Bachman's sparrow, pine warbler, Carolina wren. As I walked I came to openings in the forest where a tree had fallen after being struck by lightning, and in those patches, new pines regenerated.

I was looking for the creature most connected with this forest, the red-cockaded woodpecker, which survives best in old-growth longleaf pines. As mature pines and extensive pine barrens have become fragmented and rare, so have colonies of the once-common bird, plummeting from perhaps five hundred thousand in a historic

range that spanned from east Texas to Florida and as far
north as Missouri, Kentucky, and Maryland, to some four
thousand five hundred, mostly in Florida.

Largely eliminated on private land, it remains on a
few national forests and military bases in the Southeast.
Although federally listed as endangered in 1970, in the
1980s the bird declined by 20 percent and it continues to
decline, despite the research of dozens studying to save
it. If forest fragmentation and conversion of natural
forests to pine plantations persist, scientists warn, recovery will not happen. This, as ecologist Bruce Means pas-
sionately writes, "is another chapter in the sorry tale
of the unbridled exploitation of our nation's natural
resources."

Where the light was brighter, I stopped and looked up
at a pine with a hole thirty feet up. I knew it immediately
to be a red-cockaded woodpecker cavity. I could tell by
the swath of white resin that had dried like a dripping
candle beneath the hole. Cavity trees, in fact, are referred to as candle trees. After excavating a cavity, the
woodpecker worries the reddish sap wells around the
opening until they ooze fresh resin that dries opalescent,
forming a scabby quagmire that helps protect the woodpecker nest from rat snakes and other predators. If a cavity tree is inactive, the sap dries, turns yellow or gray; it
can be reactivated by the bird's pecking at resin wells.

Unlike most woodpeckers, which nest in dead trees,
called snags, red-cockadeds excavate nesting holes
twenty to fifty feet above ground in live Southern pines.
They will use several species of pine—loblolly, shortleaf,
and slash at least seventy years old and longleaf over
ninety. Cavity building is slow, hard work; whittling chip

by chip, this woodpecker may spend over a year and sometimes several years on its cavity. For this reason, red-cockaded woodpeckers prefer to nest in old-growth long-leaf infected with red-heart, which does not kill the tree but does make the heartwood softer. Once complete, the cavity is not readily surrendered nor abandoned. Woodpeckers may occupy a cavity for decades if they can successfully deter the pileated woodpeckers, fox squirrels, and wood ducks, who would enlarge the hollow and move in. Once enlarged, cavities are rarely used by red-cockadeds. In a typical colony, some cavities are under construction, some are active, and some are abandoned.

A family group, or clan, consists of a breeding pair and sometimes one or more helpers, typically young males that don't leave home. In this social system, two to seven of the birds live together in clusters of cavity trees—colonies—within an area of ten to twenty acres in the forest, where they defend territory and communally raise young. When young females (and some of the males) go forth to seek unmated birds, they prefer not to travel far, usually only a couple of miles. Red-cockaded woodpeckers are territorial, highly allegiant to place. They do not migrate. Mating birds bond for life, until one of the pair dies. And for such petite birds they're long-lived, often surviving for eight years or more.

The birds roost one to a cavity at night. If holes are in short supply, they will roost in scars in pine trees, cavities between limbs, or in holes in dead trees.

Red-cockaded woodpeckers are sociable and gossip with clan members a spell before beginning the day's harvest of roaches, ants, butterfly larvae, and spiders, found by scaling bark and digging into limbs. Food supply on pines is limited, so a clan requires a foraging territory of one hundred to one thousand acres, depending on the quality of habitat. On occasion they eat black-gum or wild cherry fruit or even enter fields to devour corn earworms. Clans forced into smaller foraging areas have difficulty raising young.

As ornithologist Todd Engstrom puts it, red-cockaded woodpeckers have three levels of need: tree, for roosting and nesting; forest, for foraging; and landscape, for exchange of clan members. Smaller clans farther apart make it difficult for single woodpeckers to find mates. When one member of a group dies, an individual from a nearby group can fill the gap, maintaining social structures. For this reason red-cockaded woodpecker researchers recommend protecting longleaf on the level of landscape.

Tall Timbers Research Station near Tallahassee, Florida, where Todd works, manages this old-growth forest I am in. Endowed by a northern industrialist and excellent naturalist, Henry Beadel, Tall Timbers is most famous for its research in fire ecology, especially in the longleaf pine ecosystem. Much of the Red Hills surrounding the station is in vast plantations kept by wealthy northerners as hunting preserves, and the intactness of the longleaf ecosystem there can be attributed to that self-serving but tremendously effective method of preservation, sport hunting by the well-off.

Scientists gloat about the ecological community surrounding Tall Timbers and consider this particular old-growth forest a blessing. A miracle. It is burned on a regular rotation, obvious from the blackened boles of trees, and here researchers study forest regeneration, fire regime, wildlife and plant diversity, methods of restoration, and so forth in longleaf pine. The research station continues to make strides in ecological research. Not long ago folks there discovered that wiregrass can be regenerated from seed, an insight that solves what was once a big problem for restorers of the pine system.

"You have to spend a lot of time in longleaf to appreciate it," Todd had said to me, his cerulean eyes gone dark. "This is my twelfth year studying it. You have to see it at different times of the day and of the year, different seasons, different weather. Then you understand what a truly extraordinary forest it is."

A velvet head paused at the cavity hole momentarily, then dashed out into the morning. Because I was below, the bird moved away posthaste, but red-cockadeds chattered nearby.

In that short glimpse, I'd seen it was just a bird, *zebra-backed,* as Peterson's field guide had promised, about seven inches long with a black cap and white cheek. *The tiny red cockade of the male is hard to see.* No flashiness. Nothing about it could be said to be high-profile or charismatic. Not spectacular. It was a working-class sort of bird, trying to make ends meet in a failing avian economy, depending on its clan, and in these ways and also in the way history binds it to place, it reminded me of my Cracker kin. The bird was special in the manner a molecule of oxygen is special to air.

I leaned against the stout trunk of a longleaf, its bark flaking nibs into my hair, wondering at the tree's age. It had to be a few hundred years. Core samples had been taken from the trees, and some were more than four hundred. Four hundred years: these trees had been alive in A.D. 1593. In every direction I looked I could see nothing but virgin grove.

Here was tree. Here was forest. Here was landscape. If left alone, it would function like the children on the soccer field, spaces closing and opening on a slow-ticking biological clock—a centuries-long game. A tree would fall and in its vacancy in the puzzle of sky, a sapling would sprout.

The sun was rising on another day: Bachman's sparrow, eastern bluebird, pine warbler, brown-headed nuthatch, yellow-breasted chat, red-headed woodpecker, eastern kingbird, common ground dove, quail. There was work to do, back in town. Walking out of the forest, I stopped at a downed tree to rest, startling a coyote that went bounding away into what seemed like eternity.

Poverty

Joke:

Q: "You know how to tell when you're in Georgia?"
A: "All the houses are on wheels and all the cars are on blocks."

My father fed the poor, especially the wayfaring and those who could never feed him in return. We lived on U.S. Highway 1, the thoroughfare from Maine to Florida in those days before interstates, and travelers up and down the East Coast passed the junkyard. It was the road the snowbirds coasted, and the one the vagabonds wore their soles out on.

One Saturday afternoon a drunk hobo stumbled into the junkyard. He was hungry, he said, and would my father spare a dollar for a hungry man?

"Where you from?" my father asked.

"I left Detroit six days ago," said the man. "Going to Tampa, Florida."

"Walking the whole way?"

"I been getting some rides," he said. The man walked with a limp and talked with a slur.

"What's your name?"

"Name's Elrod."

"Elrod, we've just finished dinner, and my wife wouldn't mind fixing you a plate."

"She wouldn't mind?"

"Not atall," Daddy said.

"You don't have anything to drink with that, do you?" the man asked.

"Ice tea," Daddy said, knowing the man hadn't meant tea at all.

Elrod had a hard time negotiating the steps. He was obviously very drunk—he reeked of bad wine—and couldn't walk straight.

Daddy steadied him by the arm and they shakily ascended. We children were hiding, peeping from out the door and behind the curtains. On the porch Daddy let go his grip and turned to open the screen door. The man reached for it and pitched forward, swiveling. The upper part of his body actually fell into the house but on the porch an awful thing happened. One of his legs fell off.

When we saw that, we scattered like partridges, heading to the bedrooms, but not before Daddy snuffed a twinkle from his hazel eyes. Mama was mad as a wet setting hen.

"Franklin Ray," she said. "You get that drunk out of here right this minute."

"Honey, he just wants something to eat," Daddy said. "We can feed him."

He collected the man from out of the doorway and helped him to the couch, where he collapsed. Daddy went back out and brought in the wooden leg, tacked with short straps. The man had not passed out.

"I didn't know you had a wooden leg," Daddy said. "Sorry about that. Want me to help you get it back on?"

"Give me a while," mumbled the hobo.

"How did you lose your leg?" Daddy said.

"Car accident," the man attempted, but his head bobbed and his words emerged garbled.

Mama was slamming around the kitchen, heaping potato salad on a plate and warming up the gravy. When she had filled a glass with iced tea, Daddy helped the man to the table. He didn't bother about washing his hands.

He sobered considerably after he ate, and they got the leg back on. Daddy drove the man to town and let him out, and he was on his way.

Sometimes a family like the Joads would break down near Baxley and get sent to the junkyard for a part. Daddy had a special place in his heart for the children that poured out of the packed car, dirty and tired. He'd go dig in the freezer for a bag or two of Easter candy he'd bought cheap a week after Easter, and he'd have Mama make them all bologna sandwiches. He'd help the man find the part he needed and get them on the road.

Daddy had a kindness that belied his brusqueness. His heart was big enough for all of us and a world besides, and he put innocent children and very old people at the center of it. It was so big that most of the time he had to seal it off and pretend it wasn't there.

One Christmas Day, an elderly man came to the house. He needed money and brought something—his gun, his pocket watch—to sell. He was a hoary man with gray hair and skin so thin it clung to his face and hands. He was riding his tractor because he didn't have a car to drive.

Daddy packed a pasteboard box full of food for him—
jars of canned vegetables, a loaf of bread, candy, a pack of
peanuts. Whatever we had a surplus of he put in the box
for the old man. The man carefully fit the box on the
three-point hitch of the tractor.

"Thank you, Mr. Frank. Thank you." That's what he
said, over and over, in his deep, low, gravelly voice.

But it had been raining, and there was a big pothole
in the driveway. When the old man went through the
bog, his tractor bounced wildly—I was watching out the
kitchen—and the box of food leaped off the tractor and
sailed into the deep puddle. The old man plunged down
on the brake and rushed back to rescue his food. Nothing
was broken. Daddy helped him dry it and box it back up
and tie it more securely to the tractor.

I heard stories from Mama and Daddy about a man
and woman in the 1940s buying a five-cent ice-cream and
eating from the same cone, or a woman in the 1950s carrying on, as tickled as if Santa Claus had come, because she
had ice. She was going to have iced tea, a luxury since
they didn't have refrigeration. Another extremely poor
family—the man dipped tar—lived in a shanty with no
furniture and few clothes. Those people slept on pine-
straw ticking.

When I was a girl, surrounded by poverty, the word
free meant something to me. It had an attraction to it. If
something was free, we could afford it. I ordered every
free recipe or cookbook available on the sides of oatmeal
boxes or bags of sugar. I sent off cereal box tops and got a
fifty-cent refund plus ten cents for the stamp I'd used. I
ordered booklets telling how to make gifts with aluminum
pie tins and instructions for making money selling *Grit*

newspapers. I clipped every ten-cent-off coupon, although we seldom bought anything but staples and store-brand at that.

We were poor but solvent and surrounded by people much poorer. The junkyard was bordered with a series of row houses that were without electricity or running water. Lizzie and Willie Miller lived in the first. Their furniture was spare and simple, and the house was always mopped and clean. Lizzie kept her grassless yard swept with a brush broom, the old Cracker style of fire prevention. They had an outhouse out back that attracted us intensely, so we would go stare across our hogwire fence into its stinking hole. From a greater distance we watched Willie go in and come out of it, and we snickered among ourselves as children will do. The squeaking of Willy's pump as he drew water could be easily heard in our yard. He kept a bucketful with a dipper in the kitchen.

Willie was a dozen years younger than his wife, although the age difference was barely discernible since Willie tottered around uncertainly, weaving and wagging his body, taking baby steps as he walked. He had survived a train accident in 1955. Daddy had been standing at the window of Granny Clyo's cafe when someone said, "There's a man lying in the street!" It was a real foggy morning, and the train was coming to a halt at the crossing. Before the engine, wadded up like a mouse before a cat and pushed along the track, was what had been a black truck but what was now nothing more than a snarled roll of black metal. The train had rammed the truck at the last street, and it had taken a block to stop.

The man in the street was dead. Daddy had run from the cafe, not toward the dead man but toward the train.

Another man was trapped inside the truck that had
rolled around him like a newspaper around a mess of
fish, but he was alive. That was Willie Miller. The men
were heading to the turpentine woods early of the morning and had not seen the train.

Willie, who had lost many of his teeth in the accident,
loved to suck on peppermint. He would suck a peppermint stick small and round and lick a point on one end,
then hold it in place against the gum with his tongue. If
any of us kids came around he'd ask, "Have I showed you
my new tooth?"

"No," we'd say, and he'd open his mouth.

"It's pointed," Steve would say. "It looks funny."

"Are you saying you don't like my new tooth?" Willie
would ask.

"I like it," Steve would reply. "But it doesn't look
right."

Then Willie would grotesquely drop the tooth, and
we would realize it was peppermint, that we had been
fooled.

"Peppermint!" we would laugh. So Willie would go inside and break a stick into pieces for us.

In the house past Willie and Lizzie's for a while lived a
couple named Cotton and Hilda. They had little money
and a slew of bedraggled barefoot children. (We were
barefoot too.) All I can remember is that we had to deliver something, maybe a piece of a carburetor Cotton
had traded, and when we approached the sagging shanty,
a skinny girl came out. She was towheaded and eating a
drumstick. The chicken was raw.

After Willie and Lizzie had to go to the nursing home,
an old woman of the night moved in. She, too, had spent

time in Milledgeville, though when she returned she
said, "Wadn't anybody crazy where I was in Milledgeville.
Everbody in there had good sense, looked like to me."
She wore a housecoat around all day and her face was red
and splotched, her legs swollen. She was visited by old
married men who parked down the road and crossed our
yard to her door, their caps turned backward.

A quarter mile away, John and Helen Hyatt lived without electricity or indoor plumbing in a dirt-floor shanty
sided with tarpaper, scraps of tin, and jetsam boards. It
was situated on a dirt road behind the junkyard in lowland the county had tried to drain. It teemed with mosquitoes, and the water that pumped from John and
Helen's shallow well erupted white from the ground, satu-
rated with clay. Still they poured it into jars to settle the
sediment and drank it, claiming to like the taste and calling it medicinal. They had no choice. For years they had
made a living by driving a mule and sled from garden to
garden around town, growing on halves, meaning the
party who owned the land and the one who supplied
labor, seeds, and fertilizer would split the produce in
half. John sold his share to stores in town until customers
began to complain of dog hairs in the greens. Now they
lived on a pittance of a government pension and a thin,
ragged garden.

They loved dogs. They had at least thirty of them at a
time, scraggly and raucous, penned behind their squat
shanty. They would not pass over a stray. Even when they
had little to nothing, the dogs ate. If the dogs needed a
mange treatment, John dipped them in a purple wash
that left them looking bruised and electrified. John and
Helen's backdoor opened into the dog pen, fashioned

out of bits of hogwire, rotting slabs of plywood, and half-pieces of rusty tin, and the dogs lived as much in the house as in the pen.

Sometimes we carried them meat scraps when we had extra and watched the dogs fight over bones tossed across the fence. No amount of money could have persuaded us to enter that fence. The dogs, as a pack, were ferocious. Ever so often a fight would break out between two or three of them, and Helen would wade into them, screeching at them to stop.

John and Helen had no transportation, but they didn't ask anybody for anything. When they needed food, they made the two-mile walk to the nearest grocery store. Helen was twenty years younger than John, and she would push a wheelbarrow in order to bring home the groceries and the sacks of dogfood. John must have been in his seventies. When he gave out on the trip, he got into the wheelbarrow and Helen pushed him home. I saw them pass the house many a time, Helen pushing John.

In the rural South, the land of longleaf pine, these were the pictures travelers remembered: tarpaper tobacco shanties; bent-over women in the cotton fields; shoeless schoolchildren; chain gangs; bathrooms for whites only; Saturday afternoon towns spangled with mule farmers in faded and patched overalls and not a dime in their pockets.

Passing through my homeland it was easy to see that Crackers, although fiercely rooted in the land and willing to defend it to death, hadn't had the means, the education, or the ease to care particularly about its natural communities. Our relationship with the land wasn't one

of give and return. The land itself has been the victim of
social dilemmas—racial injustice, lack of education, and
dire poverty. It was overtilled; eroded; cut; littered; polluted; treated as a commodity, sometimes the only one,
and not as a living thing. Most people worried about getting by, and when getting by meant using the land, we
used it. When getting by meant ignoring the land, we ignored it.

The Keystone

One July day after I was grown I stopped at the fruit stand by the railroad track in Baxley to buy tangerines.

"Are these grapes good?" I inquired of the fruit-stand owner, a beefy, red-faced man who planted greens and onions on the strip of curb.

"Good," he said. "In fact, I just fed some to my turtle, trying to get him to eat."

"Turtle? What kind?"

"Go look in that wagon yonder," he said.

Inside the red metal wagon with foot-high sides, used for hauling produce, was a huge gopher tortoise. I didn't recognize it immediately because it had been recently spray-painted gold and silver, gold on its carapace and silver on the finely notched head. It was drooling profusely from the nose and mouth, scrambling relentlessly from side to side in the four-foot wagon. An intense midday sun beat down on it.

It looked like a circus animal, painted like that. No telling how old it was—fifteen, twenty years old—and because it was an elder in terms of gopher tortoises its predicament seemed even more an abomination.

"Who painted it?" I asked the peddler.

"I did."

"When?" I asked.

"I just did it," he said. "I bought paint for my tennis shoes and decided to paint him too."

"How long have you had him?"

"Since yesterday."

"Where did you get him?

"Right there." He pointed toward Main Street. "I saw a van stop in the middle of the street and put a package on that strip of ground beside the railroad track. I thought it might be a bomb or something, somebody trying to blow up the train, and with me right here, they'd blow me up too. So I went over to see what they put out and it was this turtle."

"It's a gopher tortoise," I said.

"Yeah. Gopher."

"Has he been eating?" I could tell it was a he because his plastron, or under shell, curved inward.

"Naw," he said.

"He's gonna die out in this sun. They go into their holes when it's this hot out."

"He don't have a burrow."

"You know, this is an endangered species, and you could get into a fair amount of trouble having one in your possession," I said. "I'll take him and go find a piece of land with burrows to put him on."

"No, I want to keep him awhile," he replied.

"You might have to pay a fine if the wildlife officer comes by."

"Wildlife officer? He won't do anything. He's a friend of mine," he said.

"He's a friend of mine too."

"I'm gonna let him go. But I want to show him a little first."

I wanted to take a picture of the spectacle, and I walked to the truck for the camera. It was out of film.

"I want a picture of it," I said. "I'll be right back."

When I got back from Winn-Dixie two blocks away, the gopher tortoise was still prowling across the hot trailer, drooling steadily. Gopher tortoises, I knew, are prone to a disease of the upper respiratory tract, which they spread among themselves by nose-to-nose contact and which is taking a severe toll on some populations. The disease appears to be worse where tortoises are concentrated. Perhaps this tortoise was already infected, or more aptly, the atoms of paint had affected his respiratory system and the toxins would kill him before the heat did, or before he starved. Either way, what was happening to him was cruel, even if he had been saved from the highway. I tried again to persuade the fruit seller to let me relocate him to the wild.

"No. I want to show him off for a day or two and then I'll let him go."

I drove to the first phone I could find and called my father, who was home. I told him about the tortoise. I said I was minutes away from calling the sheriff, the wildlife officer, or whomever else I could find who might be sympathetic. I'd call Washington if I had to.

"Hold off on that," he said. "I'll go see what I can do." The fruit seller was a friend of his, and at Daddy's urging he agreed to release the tortoise at 6 P.M., still some hours away. Daddy called me right back. Because I couldn't locate the wildlife officer, that was what happened. Daddy carried the circus tortoise to a forest he owns. He brought his shovel with him, dug the tortoise a burrow two feet long and deposited it in its new hole. There it crouched as long as Daddy watched, but the next day when he returned to check on it, it had not deepened its manmade hole, but had moved on to make a new life for itself.

Of plants and animals native to the longleaf pine barren, the gopher tortoise may be most crucial, in the same way the keystone, or upper central stone in an arch, is thought to be most important in holding the other stones in place. The tortoise is central in holding the ecosystem together.

An ancient tortoise of great tolerance, it lives in a burrow in sandhills, flatwoods, and other upland habitat, sharing its hole with more than three hundred species of vertebrates and arthropods. Among these commensals (meaning organisms that live in close association, one benefited by the relationship and the other—in this case the tortoise—unaffected) are the eastern diamondback rattlesnake, gopher frog, opossum, rabbit, Florida mouse, skunk, armadillo, lizard, and gopher cricket. Three kinds of scarab beetles that are candidates for federal listing use tortoise burrows, as do tineid moths, whose larvae feed on the tortoise's fecal pellets as well as decaying plant material. Tortoise burrows provide refuges from cold, heat, dryness, and predators, and are especially important during periodic forest fires, when the tortoise and its houseguests escape to the hole in the ground to wait the inferno's passing. Even flushed quail run into them.

A gopher tortoise can live up to fifty years, although they take a long time to mature. Females reach adulthood at ten to fifteen years of age. At the northern end of the range, south Georgia, they may require as much as twenty-one years. Mature, they weigh about as many pounds as they are old, and the diameter of the shell maxes out at twelve or thirteen inches.

Gopher tortoises feed on low-growing, sun-loving plants like wiregrass, broadleaf grasses, and legumes. They eat prickly pear cactus, blackberries, pawpaw, saw palmetto berries, and other fruits in season. Besides unsnarled and open vegetation, gopher tortoises require open sunny spots for laying eggs and dry, sandy soil for digging, since the burrows can be up to forty feet in length and ten feet deep. The record burrow was forty-seven and a half feet long.

The life of a gopher tortoise revolves around its burrow, although it can occupy more than one. Males use an average of four to nine burrows, while females use only two to four.

Their homes, typically eight or ten feet below the surface of the ground, are long, straight, and unbranched. The burrow opening corresponds to the size of their shells—the burrow is as wide as the length of the shell so the tortoise can turn around at any point—and slopes gently downward, usually with a turn. It is too dark and narrow to see far inside the burrow. An apron of weedless sand, marked by the tracks of animals and the drag of tortoise plastron, hems the opening of most burrows.

Tortoises mate in spring after elaborate courtships that involve visits by head-bobbing males to female burrows. The female decides when she's interested. Several weeks after mating, female tortoises deposit three to fifteen eggs in the sand mounds in front of their burrows or some other nearby sand flat. Incubation depends on climate and varies from seventy to one hundred days up the length of the tortoise range. Usually the female lays one clutch per year, but often raccoons, foxes, armadillos, skunks, and fire ants raid the eggs, so that only one nest in ten years survives.

After hatching, young tortoises either join their mother in her burrow or dig their own small version nearby. Many are eaten by predators until they get too big to be swallowed, at which time their enemies are few—humans, dogs, raccoons.

Their ancestors were one of at least twenty-three species of land tortoises that originated in western North America some sixty million years ago. Along with scrub jays and burrowing owls, they were part of a savanna fauna that migrated to the Southeast. Of four species of land tortoises remaining in North America, gopher tortoises are unique in their occurrence east of the Mississippi River. Some people, because of the inhospitable climate, refer to gopher tortoise habitat as the "southeastern desert."

During winter, tortoises hibernate, although on warm afternoons they trudge to the surface to sun on the patios of their burrows. During warmer months they stay underground during the heat of the day, coming out at dawn and dusk to feed.

The gopher tortoise has been broadsided by the absence of regenerative burning in pine forests. Because it relies upon herbaceous plants for food, it is confounded by dense understory vegetation—gallberry, blackberry brambles, sumac, turkey oak—which take over in the absence of fire. Food for tortoises becomes scarce.

Because slash and loblolly pines cannot tolerate fire,

not only do they collapse food supplies, but the dense canopies that accompany pine monocultures whittle down sunny spots female tortoises need for nesting. Fire is vital in maintaining native ecosystems—most commonly longleaf pine sandhills—where gopher tortoises live.

Until it became protected by Georgia in 1977 and by Florida in 1988, the tortoise was hunted openly for food. It was a reliable source of meat during the Depression, thus the reference "Hoover chicken." In 1989, the gopher tortoise was designated the state reptile of Georgia and has been listed as a threatened species by the state since 1992.

Logging, development, and conversion to agricultural lands also have spelled its doom by destroying and fragmenting natural landscapes. Tortoises have been called the most relocated reptile in America, since developers often covet lands they inhabit. Tortoises are killed crossing highways or buried alive by heavy machinery, although studies have found tortoises able to dig out of burrows collapsed by tractors even in soil with a high clay content.

A tortoise shares its humble dwelling with over three hundred species of fauna and is, increasingly, homeless. Many of these three hundred may be doomed along with the gopher tortoise if we continue to wipe out its domain.

Beulahland

One crow sorrow, two crows joy;
three crows a wedding, four crows a boy;
five crows silver, six crows gold;
seven crows a secret never to be told.

My grandmama Beulah loved birds. For years she was the only person I knew who kept bird feeders, homemade ones Uncle Percy fashioned. She had lived on the farm with Granddaddy Arthur all her married life, but he died of cancer when I was five and I barely remember him. He and Grandmama raised seven children on the farm in two houses, one that burned down when a spark caught the wood shingles afire, before Mama was even born, and the one I knew and that still stands, which they built to replace the other. Uncle Percy was a bachelor and remained at home.

Grandmama was a small woman, hardly more than five feet, and with age she shrunk and bowed at the back. Her skin was soft and loose, and her face wrinkled in a beautiful way that showed she had always liked to smile. Her eyes, behind silver glasses that matched the soft halo of her hair, had life in them.

"Won't you look at this purty little yellow bird," I'd hear Grandmama say. Or red bird. Or blue. On many a Saturday all four of us children would go spend the day with her and those days at the farm would be long and slow. She was an excellent country cook, and when she heard we were coming, she took creamed corn out of the freezer. It was my favorite. A pack of corn would be thawing in the sink when I got there. My grandmother got as much pleasure out of me eating it as I did. At the end of a meal she'd know how many helpings I'd taken and wouldn't hesitate to spread the word.

Her sweet corn was the best I've ever tasted. She and Uncle Percy used to grow it in their garden. They picked and shucked it, and Grandmama froze it. I've been around during the whole operation and have helped Grandmama slice off row after row of warm, tender kernels into big wash pans, then scrape the white milk from the cobs afterward.

When we sat down to eat, the table would be covered with bowls and plates of food: fresh, sliced tomatoes; fried cornbread; maybe biscuits too; fried ham; peas or butterbeans; rice; okra; two or three kinds of jellies; part of a roast; potato salad; apple salad; all served with the sweetest tea imaginable.

One noon, as we cleared up after dinner, Grandma came back through the screen door from throwing scraps to the dog. "Listen at that," she said, in her sweet and royal way. "That big old woodpecker is just a'laughing."

We liked going to Grandmama's house. She had more money than we did and kept candy corn or peppermints in glass-lidded dishes and maintained a canister of cookies. Her house was big and breezy and quiet, especially after dinner, when Uncle Percy took his nap. We were always on our best behavior at Grandmama's, because she expected that and because we wanted to be allowed to return. In the mornings we helped her around the house. I always dusted the furniture. She bought furniture polish, something we couldn't afford, and I meticulously sprayed every chest and chair and table until it shined. Sometimes I polished the silver while Kay washed and rolled Grandmama's hair.

Her name was Beulah Mae Miller Branch. There was a juke joint just across the river called Beulahland, surely run by someone named Beulah, but Daddy explained that *Beulahland* meant heaven. Grandmama's name fit her then. She was easygoing and accepting, and slow to anger. She maintained the distance often required between adults and children, so we had to say "Yes, ma'am" and "No, ma'am " to her as we did to every adult and to obey her. But she disagreed with many of Daddy's ideas and fussed that he wouldn't let us do this or that. She wanted us to come to church with her. "Hogwash," she'd say when we weren't allowed, and if we needed to talk, she'd listen. Her indignation was comfort enough.

Grandmama lived into her 90s among all the stories of her life. Like this one:

Uncle Percy had brought home a magnolia tree from the swamp when he was just a shirttail boy. "He was no bigger than the tree," Grandmama said. Granddaddy said, "Go ahead and plant it, Son, but it won't like it here in the sun. Magnolias love the deep woods. You'll never set under the shade of that tree."

Uncle Percy planted it at the southeast corner of the

house and later added a maple and a holly to make a trilogy of native trees.

"Just look at it now," Grandmama said. "Fifty feet tall, I bet. And Percy was just a little thing when he brought it up."

Even Granddaddy got to sit under its shade before he died.

Granddaddy transplanted a coral vine from the swamp outside the kitchen door purportedly for Grandmama, although he loved flowers almost as much as she. It's a coral honeysuckle, whose round leaves surround the vine when they're young, like a skirt on a ballerina. The flowers are long tubes, coral-colored throats pirouetting, and hummingbirds think they're God's answer to white bread. I would walk through Grandmama's yard examining the plants, most dug from the woods: Grancy greybeard, spider lily, flame azalea, sweet shrub. Always something was blooming.

About ten different unpainted outbuildings, some of log, flanked the house, including the boiler shelter, where cane juice was reduced in a huge vat to syrup in the old days; the corncrib, used for storing ears of dry field corn; the car shelter; the shop, where Uncle Percy kept the Farmall A and his Harley Davidson; the tool shed; and the outhouse, which still stood even though indoor plumbing had arrived before Mama left home.

The old smokehouse, where they smoked their fresh meat when Granddaddy was still living and it made sense, was now used as a pantry for Grandmama's canned jars. You could buy meat cheaper and easier in Piggly Wiggly, where Grandmama shopped every Friday, rain or shine, than you could produce it.

Between the outbuildings grew huge water oaks where I would play in the jade moss that grew around the buttresses of their bases. I would set tables for hand-me-down Barbies, filling acorn cups with water on brickbat tables. The moss was cool carpet, here a nook for the kitchen, a space between roots for a living room there. It was another world, one of the mind, and in that world the trees were home.

One Saturday Grandmama yelled from the chicken coop for Uncle Percy. I knew immediately she had spotted a snake. The land where I come from is crawling with snakes, yet the sight of one triggers a cold irrational panic.

Snakes were the lowliest of creatures, condemned by God to a life spent belly to ground. One unlucky enough to reveal itself was a dead snake—nobody cared whether it was venomous or not. If a snake crossed the road, you ran it over, pulling back and forth until it was unmistakably dead. If you came upon one sunning itself harmlessly in the woods, you shot it or beat it lifeless with a stick. If you discovered one in a pea patch, you chopped it to death with a hoe.

A rattlesnake over four feet long earned your picture in the *Baxley News-Banner,* the town weekly, and when the paper came out on Thursday you were alongside Mr. Purdom with his prize roses or a farmer with a tomato big enough to slice with a crosscut saw.

Snakes seem to have an insatiable hunger for chicken eggs and chicks. Grandmama was always finding them in her coop when she went to collect eggs, bulging from having just fed. On this day a black snake curled in one of the hay-lined boxes, and its body, big around as Grandmama's wrist, swelled out oddly in two places. Uncle Percy chopped the snake's head off with a shovel and then lopped it again and squeezed out two eggs. One was my grandmama's glass egg, used to fool chickens into laying. She scalded it in a pot of boiling water, dried it off, and carried it ceremoniously back out to the henhouse.

She washed the other egg, a regular hen egg that hadn't burst either in the swallowing or the squeezing out, and cooked a cake with it.

Grandmama had an enviable collection of recipe books, some merely composition books pasted full, kept in a dresser drawer. She clipped every recipe that interested her and tested a lot of them. As my sister and I got older, we copied recipes for our hope chests on those days at her house.

My sister took to cooking from the time she was ten or eleven and, by the time she was a young teenager, baked cakes with layers thin as paper. One, once, had seventeen—a dabble of batter spread thin in a cake pan, baked while you held your breath. A cake like that took infinite patience.

I loved recipes but couldn't toast bread. I attempted peanut butter cookies one afternoon to surprise Mama, who was helping Daddy with some project or other, earnestly following the directions and laboring over them, finally sliding a panful into the 350-degree oven. I could hear our floppy basketball striking the plywood backboard nailed on the pole that strung electricity to the shop. The goal was rusty and had no net; we'd found it in some junk and put it up. I had eight minutes before the cookies would be done. Time enough for a few dribbles and tosses.

"I'll play too," I yelled, running down the back steps. Next thing I knew Mama had stuck her head out the backdoor.

"Janisse! Whatever you put in the oven is burned to a crisp."

Inside, the kitchen billowed with stinking smoke. How had I forgotten the cookies? Had it been twenty minutes?

The cookies were black disks, hard as money. I scraped them in the scrap bucket, for swilling the sheep, and scrubbed the baking sheet. There was raw dough left so I tried again, promising myself I wouldn't leave the house this time. I would stick right by the stove. I oiled the pan from the metal pitcher Mama strained fry-oil in, left over from frying chicken or corn dodgers, and spooned another baker of cookies.

This time I decided to wash dishes while the cookies baked. Three minutes. Four. Then the telephone rang. It was my friend Leah wanting to know if I'd done the math assignment, and I hadn't, so I grabbed my math book to help her. Before long I smelled new smoke.

I squealed. "The cookies are burning. I'll call you back," but it was too late. They were again unsalvagable.

It was years before Mama let me attempt baking, but I figured I'd marry one day, and I'd have to cook—all women cooked and men did not—so I'd need recipes. Mama still has old recipes, handwritten on now-yellow index cards, that she copied from Grandmama's books almost forty years ago when she was a bride. Sweet Potato

Souffle. Meat Loaf. Orange Congealed Salad. Pound
Cake. Ambrosia.

BUTTERMILK PIE
1 cup buttermilk
1 stick margarine or butter (melted)
1 tbsp. flour
3 eggs
½ cup milk
1 ½ cups sugar

Beat egg until light colored, add sugar. Mix well.
Stir in melted margarine, flour, and milk. Pour into
unbaked pie shell and bake for 1 hour.

Especially delicious was my grandmother's biscuit
pudding. I was about fifteen when I asked her for the
recipe.

"Honey, I don't have a recipe for that," she said, "I just
make it."

"Tell me how, and I'll write it down."

"Well," she said, "Crumble up three or four cold biscuits, whatever you have. Pour sweet milk over them
until they're covered and let them soak until they're soft.
Sweeten to taste with sugar. Add eggs, one for each biscuit,
and flavoring. Grease your pan and bake it about 350. The
biscuit pudding is solid when it's done. I make a good
chocolate sauce to go with mine, but you don't have to."

We had free run of the farm, and Saturday afternoons we
followed cow trails through well-grazed woods, crossing
the barbwire fence (which I thought was "Bobwire" for
years, because of its pronunciation) into the pasture and
winding toward the watering hole, a place excavated from

red clay where the cows went to drink. We'd fish or look for snakes, imagining we were Indians. We thought the clay piles on either side of the pit were Indian mounds, for the Lower Creeks had been mound builders, and it was years later we learned the mounds were formed by fill dirt from the watering hole.

But Indians had roamed there. When Mama was little, they'd found arrowhead points throughout the fields.

Sometimes we went down to the creek where Mama almost drowned when she was little. Until I was in high school a small wooden bridge—no side railings—crossed the creek. The vegetation grew thick all around, and below the bridge, the creek was mysterious and favored. We'd crouch on the bridge and peer through its holes at the brown water.

Other times Uncle Percy would make tractors for us out of a wooden spool, a spoke of crayon, a rubber band, and a stick. He even carved grips around the ends of the spools to resemble tire treads. The rubber band was threaded through the spool's center, held at one end by the crayon and the other by the stick or pencil. When the stick was wound, the rubber band tightened enough to drive the spool along the ground when it was released. Sometimes Uncle Percy made toys out of palmettos.

If he was working, we would hide from him when we saw his car drive up. He always found us. Once a couple of us thought we'd trick him and hid behind the open front door, not knowing a mirror against the far wall reflected us waiting there. He stepped inside the living room, pulled the door back a peep, and said, "Got you."

"How did you know we were there?" we wanted to know.

He grinned and pointed at the mirror.

Grandma took the local newspaper and also the
Savannah paper, which had a comic section on Sundays.
She kept the old newspapers in a stack in her bedroom,
where we devoured "Charlie Brown" and "Dennis the
Menace."

Although I was this junkyard daughter, it was easy
for me to identify with the country, its beauty, its nor-
malcy. I loved the things a farm meant: shelling peas,
making plum jelly, cutting corn off the cob. In spring at
Grandmama's the crabapples came into season; in summer the peaches and the garden; and in fall the grapes,
pecans, and pears. It was a shameless abundance. By
Saturday morning, if Grandmama hadn't called us, we
called her.

After the supper dishes were done, Grandmama
would go rock on the front porch. She was resting. We
were waiting for her to turn on the television. She'd stare
out into the yard and talk to us. She loved Spanish moss
and had her husband bring it out of the woods so she
could hang it from the trees. The crepe myrtles were full
of it, and through the drapery of moss, she watched the
world pass, what little of it there was. Few cars drove the
road, and when one came by, usually Grandmama knew
who it belonged to.

She loved flowers too, and the edge of her porch,
outside the screen, was lined with impatiens and gerani-
ums and Boston ferns in clay pots. With a whir, a ruby-
throated hummingbird no bigger than a fat crayon would
flash among the paper faces of blooms.

"Bless gracious," Grandmama would croon. "You're a
fine little fellow." Then it would whir away.

Once Grandmama caught a ruby-throated hummingbird. She was kneeling quietly nearby when it entered
an amaryllis, deep throated and crimson. Reaching out,
Grandmama easily closed the scarlet trumpet of petal, as
if drawing tight a money bag. Carefully she reached into
the blossom and drew forth the tiny hummingbird, a
feathered jewel, emerald and black in her hand. The garden light caught its magnificent red throat. The moment
she touched it, however, it went limp, so she thought she
had scared it to death. She went to the water shelf with
the holy brilliance so light in her hand, and lifeless, that
she could feel no weight, only softness. She dipped her
fingers in a jar of water there and dripped a few drops on
the hummingbird. Suddenly the little bird roused himself and, with a zip so quick the eye could barely catch it,
disappeared into tree branches.

Finally the hour arrived to watch *Lawrence Welk*
and *Wheel of Fortune,* followed by wrestling. That was
Grandmama's favorite. She'd start yelling, jumping up
and down, cheering for one or the other of the beefy,
sweaty men in bodysuits who were pounding each other
and slamming into the ring with awful thuds and yelps.
Late evening my parents would arrive for us. Luckily the
farm lay in such a way that you could see a car turn the
corner of the field a half mile up the dirt road, and by the
time it pulled into the yard, the room would not be glow-
ing blue. We'd be sitting quietly in Grandmama's living
room as if we'd been sitting so all evening, talking together, and when my parents walked in we'd be glad to
see them. No one would say a word about television.

Sometimes they'd come the back way, past the church
and across the wooden bridge and would be in the

driveway under the pecan tree before we realized it. Someone would hiss, "Get the TV. Mama and Daddy are here."

I knew how well Grandmama guarded our secret, and I wondered how many other secrets she kept, and why.

Indigo Snake

One afternoon during a visit home my friend Milton and
I were driving eastward along Altamaha School Road,
making our way toward a river bluff. Milton is a farmer
and naturalist from Osierfield, some two hours away
from Baxley, and we love to tramp the woods together. It
happened that at the same time we spotted a snake at
highway's edge and yelled "Indigo!" in one voice. I don't
know how either of us knew what species it was, because it
had barely edged onto the highway to cross.

Milton braked dangerously and left the road at break-
neck speed, then jammed the transmission into reverse.
He jumped out before the truck had fully stopped and
started running back, scouring the ditch for the snake.
He spotted it squiggling like a tiny river back through the
weeds toward the wild undergrowth. Indigos are shy. Although sixty-eight years old at the time, he dashed down
the steep embankment and cut the snake off at the wood's
edge, diving and coming up triumphant.

"Whoooo-weee," he crowed. It was a moment that
made history.

I'd never seen an indigo in the wild—they're already

that much in trouble, mostly because they're so docile that they've been overcollected for pets—but I hadn't given up hope of seeing one. Milton had told me that growing up he had seen them wrapped around "winsome maidens in carnival pits or hootchy-kootchy sideshows." The women would sit cross-legged on the floor of a tent, stroking and talking to the reptile they held, sometimes kissing it on the nose. Supposedly the snake had been charmed, but really it was simply benign, "almost never known to strike and bite."

The indigo is the largest North American snake, sometimes reaching eight and a half feet. Indigos feed on frogs, lizards, small mammals, birds, and other snakes by immobilizing prey with their jaws. Federally listed as an endangered species, they are vanishing in the wild, too, because of destruction of gopher tortoise habitat—the burrow is the indigo's retreat. They occur from south Georgia to the Florida Keys.

The indigo snake was the most beautiful thing you'd ever want to see. Milton handed her to me, and I let her wrap around my body. She was over four feet long, black as a bucket of radish seeds and mild as Talking Creek. She never stopped moving, passing between my hands, coiling around my arm, nosing between my breasts. I draped her over my shoulders like a midnight blue scarf and laughed as she circled my waist. I knew this might be the only wild indigo I would ever see. I was careful with her, not only because she was precious, but because I knew that if she sensed fear or became afraid she could inflict a nasty bite.

Milton had told me a funny story about catching an eight-foot-long indigo snake on a sand ridge the family

owned on the east side of Seventeen Mile Creek. It was
January of 1962 and he, along with a couple of hands,
was clearing the roads with bush hacks. He wandered off
alone, exploring the sand ridge, as he later told, when
he "came in view of a pile of snake that appeared to be
enough snake to fill a #3 washtub." Although it was winter, the snake had emerged from a gopher hole to sun.
Milton collected the snake, thinking to offer it as a live
specimen to a professor he knew at Mercer University,
the head of the biology department. At home he made a
pine box and attached the lid with nails nailed only halfway down, thinking he'd give the snake water the next
morning.

"Lo and behold," his story goes, "the next morning I
awoke to find an empty snake box. The huge reptile had
forced the pine planks clean off the box and escaped
into our home. Wife Mary had a few-months-old baby girl
at the time and wailed, 'That huge snake will swallow my
baby.' I knew this was impossible but couldn't convince
her.

"We all turned the home upside down for several days
in search of the snake, without success. I was certain it
had not escaped the house.

"One morning early, while we were eating breakfast,
my peripheral vision caught a swift darting motion from
behind a large upright freezer. Here was our snake. She
was coiled in and out of the heat-dispensing coils on the
back of the freezer, which backed up to a closed window.
The freezer had recently been loaded with over six hundred pounds of a beef we had just killed on the farm, and
I hated to think of unloading and reloading all that meat,
so decided the best method of recapture of the snake was

to take out the window casing from the outside and remove the lower window. This took some time and effort, and I had three pairs of eyes watching from inside the house to be sure the snake didn't move to another hiding place.

"She had herself wrapped around and in and out of the coils of the freezer, probably seeking warmth, and it took some time to get her to turn loose and come out. This was accomplished, the snake again put in her shipping box and this time the lid was securely nailed down.

"After affixing the address on the box I added 'Live Snake' in big letters. Railway Express agencies prided themselves on shipping anything, but I thought it prudent to leave the snake box in the pickup, enter the freight agent's office, and tell him what I wanted to ship. He says loudly, 'A live snake?'

" 'Yes, sir.'

"The agent said, 'Boy, don't bring that thing any closer in here. Push my scales outside on the loading platform, weigh the box, and I'll give you a label to attach to it.' "

Indigos mate from November to February (although females have been known to store sperm for as long as three years), and Milton's eight-foot snake turned out to be a gravid female. The professor hatched her eggs. If she had remained on the sand ridge, sometime between April and May she would have laid five to twelve eggs in and under logs, as well as in the burrow. They would have hatched between late July and October. In one year the young would grow to four feet long.

"The indigo was handled by hundreds of students over a period of years," Milton said, "which ends the story except for the Osierfield farm portion. Uncle Rob Dixon and Mister Fred Deese saw to starting snake stories in the turpentine quarters where over forty families lived. The last version I heard was that the snake was as long as a pickup and weighed over one hundred pounds!"

Mama

My mother was a beauty queen, daughter of a farmer,
pure as tupelo honey, the next to the last child of seven.
I think she entered the world to define long-suffering.
As was the case with most young women of the fifties, she
had no thoughts of her own ambitions. Woman came
from man, intended as a helpmate for him, and her job
was to be a good wife and a good mother. If beauty, patience, and kindness were the criteria needed to make a
home and rear a family, Lee Ada would excel. And did.

She graduated from high school and enrolled at
the business school in downtown Baxley, where she intended to learn secretarial skills in order to find a job in
Jacksonville, as her father, Arthur, wished. There she met
Franklin. She was eighteen; he, twenty. From the start her
parents did not approve of him and were vocal in their
condemnation. Franklin was a charmer, tall and handsome, forehead wide as a broadax, but he came from bad
blood. He was allowed to see Mama two ways, by sitting
evenings with the family in the living room or by riding to
church in the backseat of their car, Beulah and Arthur in
the front, Franklin and Lee Ada in the back.

When Daddy asked Mama to marry him, she said yes, giddily, her heart singing. They were of age, with no need to ask Grandpa for Mama's hand although it would have been no use, for his answer would have been a curt no.

On a searing day in July my parents eloped. They arranged that Daddy would pull up to the house at ten in the morning, when Grandpa, siblings still living at home, and Mr. Sealey, the hired man who lived with them, would be deep in fieldwork. Mama would be ready. The fields of tobacco released a sweetness of nicotine in the air, until just breathing it made them heady, and the walls of woods at the edges of the fields were profound and green. They went first to the Spring Branch preacher, who was fixing a piece of machinery.

"I cain't," he told them. "Arthur would kill me."

"Preacher Braswell," said Daddy. "We're eighteen and both of age. We're going to get married today. If you don't do it, somebody else will."

"Not in the church," the preacher said. "Let me change into good clothes, and I'll meet you here in a half hour."

Mama and Daddy waited nervously. She was wearing a calf-length white dress with silver bows and her pin-curled hair framed her lovely face. She was the prettiest sight my father had ever beheld. By now Mama would've been missed, although more likely she had been glimpsed leaving the house or seen in the front seat of Franklin's car dressed fit to kill when it turned into the road. What would they be doing so surreptitiously, ignoring the rules of Arthur's household, except eloping? News travels fast and mysteriously in the country.

When the preacher reappeared he had his Bible with him, and he led the hopefuls beneath the eaves of nearby

wooden Hamilton School, ducking into shade that
spilled from the roof. The day was unbearably hot. Flecks
of sweat glistened on the preacher's upper lip, and Daddy
took out his handkerchief to mop his angular face. His
hair was slicked back and his thin frame did not fill his
suit. Below their feet the drip line etched a visible line of
hollows, like those of ant lions, in the bare sand. The line
divided the world as surely as the desire to control our
world divides us from the wild, and Mama and Daddy
stood on one side of the line.

Afterward, they drove the two miles home to Mama's
farmhouse to collect her things. Arthur was coming into
the yard, stepping toward the outside sink where he
would wash up for dinner. When he heard the car, he
turned and leveled a long, hard look at it. His daughter
alighted and skirted the fenced flower garden, where
Beulah's bottlebrush bounced with sulphur butterflies,
to approach him. He waited.

"Daddy," she said quietly, sweetly. "You want me to
come help you crop tobacco next week?" Tobacco was
the Branch's cash crop.

Silence stretched between them. She sensed the anger
and hurt lying big before him and knew that he did not
mean to imply by what he said that she had eloped to
prove a point. He spoke one harsh line. Even in his bit-
terness, a note of resignation laced his voice.

"You got an ax to grind, go grind it," he said.

Sometimes I was jealous of my parents' love. What lay
between them even their children could not approach,
though we had come from them. Through the vast and
empty acres of my child's mind, the extent of my world, I

searched for them, knowing I would find them together, side by side. Truly, theirs was a good marriage. Once, in the locked safe where Daddy kept lost teeth, buffalo nickels and silver dollars, special drawings, and the first poem I ever wrote, I discovered a love note he'd written to my mother.

"Yes, Lee Ada, I will love you forever," it said. Their love was a private thing that belonged to a hollow in a soft and flowering meadow, a place I could neither find nor enter.

"She used to ask me so often if I still loved her," my father grinned, "that I wrote down the answer. So she wouldn't forget."

On one of Mama's birthdays, Daddy gave her a home-made knife on whose crude handle he engraved, "If the going gets tough, baby, cut your way out." When she and Daddy repeated after Preacher Braswell Carter under the eaves of Hamilton School "til death do us part," they meant it. Only death would separate them.

My mother's answer to Daddy is on a quilt she made, a kaleidoscope of beautiful log cabins pieced of scraps of forest green, pink, and blue, at the center of each a square of bandanna red. The polyester backing is red as well, and in the middle of the quilt's back is a white heart appliqué. On this heart Mama embroidered, years after they married, "I love you, Franklin."

The last chapter of Proverbs describes Lee Ada better than anything I could attempt: *Who can find a virtuous woman? for her price is far above rubies.* My mother did not have to work at virtue—she was born to it—angelic, simple, kind. As I reached womanhood, when I was first hot for equality, justice, and freedom, virtue meant no more to me than cow dung. I was impatient with my mother's refusal to assert herself. Only years later did I appreciate her wisdom, her steadfastness. Her life is one long gift to those she bore and loves. One long back-breaking gift.

She riseth also while it is yet night, and giveth meat to her household, and a portion to her maidens. Often after her husband and children were abed, I would wake and hear the washer spinning or the dryer tumbling or chairs being scraped across the kitchen floor and know that Mama was yet up, sorting and folding laundry, sweeping and mopping the kitchen.

She is the hardest worker I know. She worked like a dog. Too hard. She should have required us to do more—if I was late getting out of bed and dressed for school and left my bed mussed, it would be made when I got home from school. Mama did all the laundry, all the washing and drying and folding and putting away. Until Kay got big enough to help, she did all the cooking. She washed the dishes and the pots and pans. Eating out was a luxury we could not afford—unfathomably extravagant. I remember eating out once—seafood—when the Kiwanis Club honored me for being Star Student. Mama's only break was Sunday nights, when we had cornflakes. She never had much money to work with, and her house was filled with simple things, but it was spotlessly clean. The house was her domain and through sheer determination she would keep it well. She drove herself to exhaustion.

She was the first one to rise in the morning, by six-thirty calling us out of sleep and laying out to cook a full breakfast, even if she and Daddy were fasting—grits, bacon, eggs, toast. Every morning. She eyed the clock and, when the school bus was due, watched for it to round the corner.

"Here it comes!" she'd alert us and hold the end door while we rushed into the days of our lives. There was no time to peck her on the cheek.

We were her existence. When the bus delivered us home, she might have homemade cinnamon buns waiting, thick with pecans and raisins, strong with cinnamon, coated with sugar icing. She might have oatmeal raisin cookies, still warm and aromatic. Or she might have been helping Daddy outdoors and have to fix us cinnamon toast, slices of white bread dotted with margarine, sprinkled with sugar and cinnamon and grilled, for we would be ravenous. *She looketh well to the ways of her household, and eateth not the bread of idleness.*

My mother was a simple cook. She prepared foods she'd been raised on, plain Southern fare—rice, gravy, sliced tomatoes, turnip greens, cornpone, grits, eggs, chicken and dumplings, pot roast, ham, field peas, lima beans, potato salad, stewed okra, pumpkin pie, salmon balls. We didn't have fancy casseroles or lasagna or spaghetti, and nobody had ever heard of a burrito or an egg roll. I didn't know what an artichoke or parsnip or kiwi or papaya was—certainly had never tasted them. We drank sweet iced tea and sometimes lemonade.

Mama made biscuits the old-fashioned way, hollowing a well in the bowl of flour and cutting the shortening in with her fingers, then pouring milk into the reservoir and stirring until she had a ball of dough. She pinched the biscuits off one by one, rolling them into small balls and pressing them out on a cookie sheet. Each biscuit bore the mark of the back side of her knuckles.

Her work was not confined to the house. Because she loved Daddy, she worked alongside him in the most strenuous of toil, hauling lumber and cinder blocks, re-building engines, hauling junk. Her skin was brown from being outside so much, and she bruised easily. *The heart of her husband doth safely trust in her, so that he shall have no need of spoil.* He was a lucky man.

The inside was her second job, and while Daddy rested or napped, she cooked and cleaned and dusted and swept and mopped. At meals she waited on him hand and foot. "Cook," he affectionately called her, "pour me another glass of tea, would you?" And she'd get up from her chair and pour it. It was her job and she did it all without complaint. She was Superwoman disguised as a chaste Cracker housewife with four children and a hus-band who smelled often as not like car grease.

Even when I was old enough to know better, some-times I was not good about helping Mama. If we had any spare time (when it rained—and thank God for each of the fifty-odd inches of annual rainfall), I wanted to read. Mama complained that's all we did. Reading was our gate into other realities—through it we could escape the hard-ships of our own days.

Our cramped public library was limited. We read every children's book they had, plus some of the adult books. For some reason I was hungry for knowledge of the pioneers, eager to understand what great hope moti-vated them to leave homelands and set off for promised lands in the dangerous American frontier. I read *Caddie Woodlawn, Little House in the Big Woods,* the biography of Daniel Boone, anything that would take me there. I cut fringes to hem an old brown shirt so that I would look like Sacagawea or Coosaponakeesa.

When my father was sick, my mother stepped into the role of head-of-household without a stumble. When he gave away our red 1965 Mustang during one of his episodes, Mama calmly called the new owner and had him return it. Another time, when Daddy decided to back out of a land-buying deal, Mama telephoned the realtor.

"Franklin's on his way to see you," she said. "He's not well. Ignore what he says. We want to buy the land."

Her decisions were swift and sure, and she kept the family afloat. Where the money came from to feed us I don't know, because she never worked an outside job.

My sister remembers needing a pair of shoes when hers wore out. "We don't have the money," Mama apologized, and when they went to bed that night, Kay sneaked Daddy's wallet, just to see what was in it. He was home at the time but not doing well. The billfold held seven dollars, which, even for 1970, wasn't much.

I honestly can't imagine how they made it during those hard years. I know Daddy tried Social Security at one point. He wasn't able to support a family, four small children. The Social Security people looked at their books and said, "Yes, our records show you've paid in, you're eligible," but they wouldn't give him back a cent.

"And if ever there was a time I needed it," he said, "that was it."

After 1971, Daddy's mental illness did not recur

except in minor irritabilities and obsessions and the inability to concentrate on more than one thing at a time. A doctor in the state hospital had taken special interest in him, and one time before he was dismissed, she called him into her office. She advised him to keep his life as stress-free as possible, not to get too worked up over anything.

"Second," she said, "make sure you and your family eat a nutritious diet that includes a wide variety of foods. Some research blames the high rate of mental illness in Georgia on poor diets."

Mama loves knickknacks and other pretty things. Dainty ceramic frogs and birds and deer cluttered every available horizontal surface of the house. She especially liked miniatures—three-inch cast-iron skillets and tea sets and thumb-sized roosters. Daddy built whatnot shelves, and Mama filled those as well. She collected porcelain bells and figurines and dolls—a little McCoy girl holding a teddy bear; a marble goose; a music box that looked like a produce stand, with minute barrels of red-speck apples and a jar of green-sliver cucumbers on the counter. Although on the farm she was not poor enough to go hungry, she had been deprived of pretty things. She craved them. She loved antiques too and displayed shiny kerosene lamps and milk churns and wooden clocks wherever she could. Each had a story. Daddy's junk deals often netted knickknacks and antiques, and Mama was as eager as he to go through boxes of stuff to see what they held. She'd wash the treasure in hot bleachy water and find a space for a miniature girl at a well or a cut-glass candy bowl.

Old dishes, especially Depression glass and real china, enchanted her, as if she could absorb the richness of years from these things. She liked to know where they had come from and who had owned them before her. She had crockery bowls from Daddy's grandmother and from her own grandmother and a vase from Nippon. She packed kitchen cabinets and her china cabinet with beautiful unmatched bowls and gold-rimmed plates and fragile teacups and saltcellars. I suppose it is safe to say the house replicated the accumulation of the junkyard, and in this way, too, my parents were well matched.

Once I read Thoreau I wanted none of it.

She is not afraid of the snow for her household: for all her household are clothed with scarlet. For years Mama sewed most of our clothes, including dress suits for the boys, complete with lapels and linings. I remember one red corduroy coat she made that caused people to call me Little Red Riding Hood. She sewed shirts for Daddy, difficult projects with their collars and facings and cuffs. When else could she have done this but in the middle of the night?

With scraps of cloth Mama stitched gorgeous quilts as her own mother had done, in all manner of color and fabric and pattern: Dresden girl, wrench, pinwheel, patchwork. Winter evenings found her cutting out colorful triangles, rectangles, and squares from cotton yard goods, stitching them together while we read. *She seeketh wool, and flax, and worketh willingly with her hands.* At first she would quilt at Grandmama's house until Daddy made a quilt frame that she set up in the big house.

At times, when most tired, Mama's long-suffering ended, and she reached the limit of her patience. Then she might yell at us for tracking dirt on the just-mopped floor or strewing something she'd already picked up. She would get angry at us for fighting among ourselves. Once I remember her chasing Dell around the table with a fly-swatter, and because she could not catch him, she threw it. Daddy was the disciplinarian, not Mama—all he had to do was touch the buckle of his leather belt to stop our horseplay. She seldom laid a hand on us. Either she told Daddy and got him to do it or fussed at us until we were good and sorry. Rather than getting angry, Mama broke down and cried. Her crying was awful. You felt as if the world might end when she cried. It was miserable to see and hear. There was nothing you could do to comfort her.

I was poor help for Mama. For one thing, feminism came early and naturally to me. Why should I have to wash dishes when the boys didn't? If Mama required them to make their beds, her load would be that much easier. Like Kay, I made mine. Mama would hear me out when I raved at her about the injustice of women's work, but she would not change the way her household operated. I told her that it was unfair. I hated being inside as much as the boys did.

"Tomboy," Mama called me.

"Tomboy," my sister told my grandmother.

Growing up, I thought my mother was beautiful and loved her desperately, but I did not want to be like her. She had given up too much—her own opinions, even— to marry a strong man and be his helpmate, though he fathered her children and provided for her family and stayed loyal to her all the days of his life. The needs and desires of family eclipsed Mama's own. Yet she is the most steadfast, generous, and honorable person I have ever known, wise in her unassuming way, and because of this, she approached sainthood. On these terms I did not want to be a saint.

Bachman's Sparrow

Another bird distressed by the diminishment of longleaf pine forests is Bachman's sparrow, small and nonmigratory, a bird so suited to open pine savannas with little to no understory that it has been unable to adapt to dense pine monoculture. Since the 1930s it has declined at a stunning rate. Bird-artist John James Audubon discovered the sparrow in 1832 while exploring near Charleston, South Carolina, and named it for a Lutheran minister he had befriended on the street and with whom he was staying, John Bachman.

Bachman's sparrow is streaked buffy-gray, with a shadowy bill and a long, dark-brown, rounded tail. It measures six inches from bill to tail tip, about the size of most sparrows. It has sometimes been referred to as "stink bird" by quail hunters, because hunting dogs can be distracted by the scent of ground-dwelling sparrows.

This letter I imagine penned in the great beyond by John James Audubon and sent to his dear friend, Reverend Bachman.

My dear John,
Remember if you will the small finch I discovered within a

*scant hour's walk of your home when we both dwelled briefly
on the good earth. Although I was myself a visitor in your
godly mansion, where hospitality and kindness were non-
pareil, neither of us understood then how thoroughly each of
us was visitor.*

*I am reminded of this passerine of late due to increased
sightings of them here in the ethereal longleaf pine groves of
our Lord. They have become quite abundant on my rambles,
which can mean only one thing: they become less so in the
world below. I welcome the blessed sparrow.*

*Consider how dull our colleagues thought it when we first
brought it to the attention of science. Its humbleness so nearly
matched your own. There only in song could it display its absolute glory— the morning's loud, clear whistle followed by a
shower of operatic notes. This from the males.*

*The dawn I introduced you to your namesake we walked
through a grassy opening in the pinelands. Such were the
slants of fresh sun through a translucence of fog that one felt
he might tug himself up into the heavens by means of one of
them. The forest, out-of-focus but immediate, was surreal, as
grand as anything ever I saw, and as you know, my short life
on earth was spent in constant search of the beauty of God's
creation.*

*That morning, if memories of the terrestrial world serve us,
we were speaking of my travels in the Florida Keys and the rap-
ture of seeing roseate spoonbills. You recited a verse to me from
Genesis, about the sixth day of creation:* And God saw everything that he had made and behold it was very good.

To think that we were loathe to leave that earth.

*One of the sparrows began to call, not unusual for them
in late summer, and I fell silent. The plaintive yet sweet notes
ended. Then, because we were close enough and listening*

intently, we heard its "whisper song," the notes it sings sotto voce.

"*Bachman's sparrow,*" *I said. You nodded humbly and smiled.*

Purely by chance we came upon a female nesting in a grassy dome she had erected over a shallow depression. The nest was shielded by a clump of palmettos, and we happened upon it by watching a bird disappear, then going to that entrance. It was not yet May and three glossy white eggs cradled in her nest. Later I found nests with four and five eggs, never more, and learned that the young hatched within a matter of ten revolutions of the planet.

Sir, not long ago I was with St. Francis and some of the angels when passenger pigeons crossed overhead. The only darkness I have witnessed here is the temporary one created by the shadows of their crowded wings. In the deafening chorus of their cries, St. Francis turned to me, eyes a'twinkle, and said he felt as if he'd died and gone to heaven. I have no reference for the length of time we stood beneath their flight.

Your sparrow seems happy here, although its presence in such abundance does not bode well for our dear ones left on earth. The creature was never common.

Not long ago, Mr. Peterson, a man who loved birds with the passion of Christ, as I did, came to be with us. I heard of his arrival and made the journey to welcome him. He recognized immediately my name when I spake it, and beams of light poured from his eyes. He could not contain his excitement.

"*John James,*" *he said. "I have wished all my life to thank you for the paintings." So overcome was he, I offered to take him to the floodplain of the river, where ivory-billed woodpeckers forage in virgin deciduous trees. When he saw them, tears issued from his eyes as a man bereaved who finds loss restored* and is suddenly aware of the profundity and vastness of his heartbreak. Ruby-throated hummingbirds hovered beneath his chin and drank the tears that spilled from his eyes.

I told him I have been unable to look upon the world of flesh of late and inquired of the state of affairs below. He described large and ever-growing cities where millions of people are crowded. Their water is not fit to drink. More and more they forget the ways of birds and snakes and butterflies, and the marks of seasons vanish. They hide from stillness and darkness.

Much of the tallgrass prairie is gone, he reported, as are the spruce-fir forests of the southern Appalachians, the bluegrass savanna of Kentucky, sedge meadows of Wisconsin. Little is left of the palouse prairie of the West and the pine rocklands of southern Florida. Only a few thousand acres of virgin old-growth longleaf pine remain. As the old longleaf savannas are logged, he said, your sparrows are unable to nest in the dense underbrush that springs up. The bird prefers older stands because the thick ground cover means safety — cleared forests are only briefly suitable, before woody undergrowth shades out the forbs, the broad-leaved flowering plants. The last he'd heard, the songbird was extirpated from Maryland; almost gone from Tennessee, Virginia, and Georgia; and in serious decline in Florida, where he had recently seen them.

I have sent a Carolina parakeet to deliver this message. I promised to show Roger the herds of bison after the Feast of Forgiveness. If you wish to accompany us, we would be grateful. I have sent an invitation to St. Francis as well.

We can pass through the longleaf on the way to the prairie. For a second time I will be able to show you the bird who so *honorably bears your name. If we are fortunate, red wolves will be calling.*

Your affectionate servant,
John James Audubon

Light

When I pick up my childhood like a picture and examine it really closely, I realize that I left home not knowing how to swim, not knowing the name of one wild bird except maybe crow, and that I couldn't identify wildflowers and trees. I knew the Dewey decimal system inside and out, could calculate the force of gravity on a ten-pound block sliding down an incline, had read Dumas and Chekhov and Brontë but couldn't tell a weasel from a warthog. I never knew a naturalist or that there was such a thing as an environmentalist.

Lucia Godfrey, who taught science to the fifth and sixth grades, was a person who nurtured my interest in the living world. One lovely spring morning I remember walking beside her on the school playground. If the grass had been greener it would have burst into song. The sun's fire smelted gold.

I had tried to play football with the boys in my class, but they said no.

"Why not?" I demanded, halting their play.

"Just because," they said, no one daring to offer a reason. I knew it wasn't only because I was a girl, for they had allowed Suzanne to play—she was no bigger than a thimble and cute, not a threat. But I had to wear dresses all the time, and that made them uncomfortable. How could a girl in a dress play football? If she got tackled, her dress might fly up and reveal her panties. They were already embarrassed and confused, since I looked like a girl but certainly didn't act like one. Tomboy. These guys played junior varsity, and they knew how to play football— me playing didn't look right.

"No," they said.

Someone passed the ball to Timmy Byrd, the fastest of all, and he headed down the field pursued by a pack of sixth grade boys and one sixth-grade girl. Without thinking I was running too, merging with the boys, then drawing ahead until I was alongside Timmy. Reaching out my long skinny arms I tackled him and brought him down to the ground with me. The earth was hard enough where we landed in a tangle of fire and muscle to skin one white leg with specks of blood. My skirt, knee-length and polyester, was dirty and grass-stained down one hip where I had landed, Timmy on top of me. He was furious.

"Nobody said you could play," he shouted. "Why did you tackle me? OK, you want to play. Here's the ball, run with it. Here!" He thrust the football at me.

I'd never outrun a mob of boys, especially one with injured pride. All of them were friends of Timmy's, and despite them all saying no, I had joined the game anyway. They stood around gasping and heaving, chests out and arms at sides, their boy faces slick with vengeful sweat. They would pummel me.

"Hell, no," I said, cursing to shock them and rescue

my dignity. Cursing got you sent to the principal's office. "Damn no," I added, and walked away.

Mrs. Godfrey had recess duty. She hadn't been long moved to town with her husband, a retired colonel, and four pearl-skinned, almost-grown daughters. She talked funny, and I liked listening to her, and I liked her alabaster skin and the pretty colors of lipstick she wore, coral and Elizabethan red and poppy, which prompted me to experiment privately with rancid lipstick found in old cars, my sister and me in front of our mirror.

I crossed the playground alone to the covered sidewalk where she stood in sunlight. She looked down at me once and smiled with her lips pressed together, but she didn't say anything. I wondered what she had seen. She had not heard me curse.

"Oh, look," she said after a while, "the pine trees are flowering."

I'd never seen a pine tree with flowers. "A pine tree has flowers?" My hand stung where I had skinned it.

"See the yellow spires?" she asked. We strolled toward the chain-link fence that separated the school yard from a housing project.

"That's male," she said. "Those are its flowers. They're called candles. They're filled with pollen, to fertilize the female flowers."

She craned her head, looking for a woman tree, and I thought how elegant she was. Even her name was exotic: Lucia. She had told me it meant light, and where she came from it was pronounced with a *ch* instead of an *s* sound. I liked the way her black hair shone as she explained that the female part of the tree was the cone.

Mrs. Godfrey was often contemplative. She went about her duties with her head tilted down, appearing to think constantly.

She talked quietly to me, gazing without judgment into my face, as if she had been marveling at pine tree reproduction for all of the tiresome recess and waiting for someone, me, to come up to her, so she could show me today's miracle. I wondered if she thought about anything except biology, if she loved her daughters the way my mother loved us, what she cooked for supper, where she had been raised. Sometimes I asked her personal questions that she answered briefly, with a yes or a no, without emotion or explanation.

She did not care that I was different. She never once asked about my home life, knowing nothing more than that I was a child hungry to learn, which she read in my face and also in the way I hovered over the microscope, drawing invisible animals that pumped and collided through single drops of ditch water we delivered to school in mayonnaise jars. I sat in the front row, close to her, as she taught parts of cells and kinds of rocks, as if this knowledge might deliver all of us from pain. Knowledge was what she knew how to give.

Out of all her science lessons, that one on the playground not only did I never forget but remember as vibrantly as if it happened last week. I learned that nature wouldn't ridicule you, would let you play. Oblivious, it went about its business without you, but it was there when you needed some gift, a bit of beauty: it would be waiting for you. All you had to do was notice. When the bell rang to head to class, I bent to pull up my socks that drooped even though I had rubber bands folded under them, just

below the knees. The ground caught my eye. Pollen
dusted the grass yellow.

I told Mrs. Godfrey about the pitcher plant in the junkyard, and she had me bring one to class. She said the
stemmed pitcher is really a leaf although it doesn't look
like a leaf at all. She said that because the plant needs
more minerals than the soil provides, since they grow in
infertile places, they found a way to utilize the nutrients
of insect bodies. They adapted in order to survive.
    Evolve. Adapt. Survive.
    In spring I brought a *Sarracenia minor* flower to Mrs.
Godfrey, because it proved the pitcher cannot be anything
but the leaf and because she was the kind of woman you
wanted to give pretty things to. She would appreciate a
pitcher plant's bloom and would know that I had walked
alone far down into the junkyard, listening for scary noises
and watching for the bad sheep, the one who would chase
you and butt you if you didn't scamper up some junk to
safety. She would know that I had not asked permission to
go, but had sneaked away after we finished loading batter-
ies my father was taking to Waycross to sell, when I should
have been inside helping my sister peel Irish potatoes for
supper.
    "My, how beautiful," she said, exactly what I wanted to
hear. She looked straight at me. Her eyes were black as
little universes. "Thank you."
    Perhaps something could have been different for me.
Certainly not adulthood, for we become our heart's desires, but childhood—could the natural world I now re-
vere have opened to me? Suppose someone had found
my father the boy and said, *If you look closely, you will find*

palmetto bugs hardly bigger than apple seeds, and their irides-
cent black shells are walking onyx. And, *A yellow-rumped*
warbler is in the wax myrtle. The eggs of fairy shrimp spread
by wind.

    Suppose. What then?

Flatwoods Salamander

It is raining. This is the second long day of dreary rain, pushed out of the West in advance of an October cold front. The tapping of water stirs an eternal urge inside the intricate and fragile bones of the flatwoods salamander. The time has come to breed.

Breeding requires a journey for which the salamanders do not prepare. Instead, when the breeding urge fires the fibers of their stringy muscles, they turn from the rain-soaked grassy uplands of longleaf forest where they have been living among a tangled field of wiregrass, pinewoods bluestem, and toothache grass. The grassland is riddled with crayfish holes and root burrows, and they have been feeding on worms and other invertebrate fauna of the soil.

They begin to crawl downslope, although in this flat country the decline is so imperceptible they can feel it only in the deepest recesses of their spines. They turn their 3½- to 5-inch bodies toward the lowlands, the place of their own hatching. They do not stop. The darkness does not stop them. Rainwater leaks through the clasp of grass and slickens their bodies, drenching the sandy dirt they hurry upon. Along their backs, gray net patterns shape-shift with the wag of small feet. Not even a fool would be out in such weather, unable to see the way ahead. Yet here I am too, riding the highway looking for them. Blindly the salamanders crawl, faithful to old processes lodged in their tiny skulls, faithful to the place of beginnings.

A picture of that place is soldered into their brains, a map to it etched with the passing of millenia. Through time, this map does not weather but stays sharp and demanding inside them.

Over tussocks and around clumps they travel, seeking ephemeral pools formed seasonally in shallow basins in the flatwoods. Here they meet mates and breed in the wet darkness. Females will oviposit up to 160 eggs in small groups under debris or on bare soil in the grassy upper zones of slow-filling ponds before making the arduous return to the uplands. In three to five weeks, the eggs will hatch into half-inch larvae with brown-and-white stripes.

In three or four months, when the pond has begun to dry up, the three-inch larvae undergo metamorphosis and resurrect from the shrinking water to become terrestrial. Adult salamanders can survive only on dry land.

For centuries, flatwoods salamanders have migrated over five hundred yards, about a third of a mile, each way to and from breeding grounds. Over time, as the pilgrimage becomes more challenging, fewer and fewer of them succeed. In the absence of fire, woody plants overtake pine savannas, and these become barriers that confuse the salamanders. Two- and four-lane roads are paved between living and breeding territories. Forests are logged;

wet areas, drained. The map inside their heads no longer matches the terrain.

Flatwoods salamanders are a lot like striped newts, another rare inhabitant of pine flatwoods. Striped newts are aquatic salamanders about four inches long, not much longer than a stick of gum. They fit in your palm. They are olive green to dark brown, with two red stripes running the length of the body.

Endemic to south Georgia and north Florida, striped newts have a complicated natural history. As orange-red efts, they live in upland sandhill, scrub, or pine flatwoods communities with a healthy herb understory. In winter, they migrate as much as 750 yards to breed exclusively in ephemeral pools, the puddles that fill during wet seasons and vanish during droughts. For striped newts, the length of time the pool is filled with water must be short enough to exclude fish, which prey on the larvae.

After hatching, small larvae follow one of two pathways, though how they decide which career to undertake is unknown. They may transform into terrestrial efts, ridding themselves of gills and fins, and claim a piece of the uplands, returning the next winter to breed if conditions, such as rainfall, are favorable. Or, striped newt larvae may remain in the pond, retaining gills and a full tail fan although they are sexually mature and breed during their first year of life. Most move into the uplands as terrestrial adults in the summer, likely returning in subsequent years to breed in the pond, where sometimes aquatic adults, fully transformed and lacking gills, are found.

Like flatwoods salamanders, populations of striped

newts, too, are diving, although they are listed as protected in both states. Since they are most common in areas maintained by regular burning, fire appears to be one of their requirements.

In 1970, ecologist Bruce Means, my teacher and friend, located the largest known population of flatwoods salamander along a three-mile slip of paved county highway in the Florida panhandle. Salamanders are easiest to find during migration, and for two years he did a census of the population. During a steady rain the night of October 9, 1971, he observed seventy-one individuals in one hour crossing the road. On that night in his field log he wrote, "If this migration lasted all night (and I suspect that it did) there were 200 to 300 adults moving across the road."

Over the course of his two-year examination in the early seventies, the rate at which he encountered flatwoods salamanders was an average of 7.9 individuals per hour, most crossing east to west.

During the late 1980s, Means casually studied the site and was shocked to encounter only 0.6 individuals per hour. Suspecting a decline, between 1990 and 1992 Means made a concerted effort to count migrating salamanders. Twenty hours of viewing time on twelve trips produced only two salamanders, or 0.1 per hour. The community was essentially extinct, a decrease of 99 percent in about ten generations of salamanders.

Means searched to find the blame. Drought was a possibility, but rainfall data revealed no trend toward drought. There appeared to be no increase in highway traffic. Acid rain did not resolve the question since breeding waters of the cypress ponds are highly acidic anyway.

On the east side of the road lay national forest, where the only change in twenty-two years was increased closure of the longleaf and native slash-pine canopy due to the exclusion of fire. The land to the west, however, was a different story. It was privately owned and had been roller-chopped, bedded, and planted with slash pine, where longleaf had originally grown. I saw this with my own eyes.

Bedding is a common, if destructive, silvicultural practice in wet soils because it elevates the newly planted trees in order to decrease inundation of their roots, creating a false environment. Bedding uproots cover plants and buries them. After bedding has been carried out, native perennials decrease pronouncedly and disturbance-loving plants, like broomsedge, dog fennel, and brambles, flourish. In time and without fire, woody plants like gallberry, myrtle-leaved holly, yellow jessamine, greenbrier, and swamp titi creep out of the swamp and march up slope. Furthermore, bedding jumbles time-ordered layers of soil.

Within a decade at Means's study site, rows of tightly packed slash pine, planted in long beds, had created a canopy of dense shade the land had never before known.

Means told me he believes physical alterations of terrain interfere with the orientation of the flatwoods salamanders. They may wander endlessly in search of home ground. If the bedded ridges lie at angles to their paths, they may be diverted away from the ponds. Or impregnated females may lay eggs in disconnected puddles that leave the larvae high and dry, without sufficient time to develop before the shallow water recedes beneath a thirsty sun. The females may oviposit on the tops of furrows, mistaking the wetter trenches for ephemeral pools and thinking the furrow is its grassy edge.

Silviculture was the only ecological change in twenty-two years Means found in this flatwoods salamander territory.

"If we are correct that slash pine silviculture is in some way responsible for the decline of what was once a large breeding migration of the flatwoods salamander on our study area," wrote Means, "then silvicultural practices may explain why the geographic distribution today appears to be fragmented and why known populations seem to be declining or extinct."

In 1998, forty-seven known communities of the rarely seen salamander were left, three-quarters in north Florida and a handful in Georgia and Alabama.

Altamaha River

In a time before memory, when I was a fat-cheeked three-month-old, my parents were motorboating with my sister and me on the Altamaha, an alluvial river 137 winding miles long and twenty million years old, formed from the Oconee and Ocmulgee rivers, which drain middle Georgia and join near Lumber City.

The Altamaha is Georgia's largest river and the second largest river basin along the Atlantic Seaboard, I'm told, exceeded in water flow only by the Chesapeake Bay system. It empties more than one hundred thousand gallons of water into the Atlantic Ocean every second. It's crossed only five times by roads and twice by rail lines and is full of snags, parts of fallen trees, many of them concealed just below the surface of the sediment-filled waters.

As the evening of which I speak had begun to slip toward darkness, my father turned upriver, the landing yet miles away. He was in a homemade boat, one he'd built following *Popular Mechanics* instructions. It was called a PM38 because supposedly you could build it in 38 hours for $38 and it would go 38 miles per hour with a 38 horsepower engine. Daddy was midstream and full throttle when the boat struck a snag that ripped a hole halfway down its hull. Within seconds the boat was submerged, but by working with the current, Daddy managed to maneuver it close to shore before it sank. I was strapped alone on an infant's life preserver, as were my mother and sister, and when the river took me, I bobbed up and down with it, spewing water, and floated to shore. Daddy swam in and all were accounted for. He picked me up and untangled me from the preserver.

"Was I crying?" I asked Mama later, as she retold the story.

"No," she said. "You seemed at home."

My father knew the Altamaha stole drunks, children, and old fisherman; every summer a couple more drowned. He considered the accident a sign. He repaired and sold the boat and quit going to the Altamaha. "I'd try not to even cross the river bridge," he said, speaking of the U.S. 1 bridge you had to cross to get to Vidalia or Lyons. The concrete structure, built in the mid-1930s, was a mile wide to compensate for floodplain, because in rainy seasons the river leaps its banks and spreads wide through the buttressed trees. As kids, we made a game of holding our breath as we crossed the bridge, craving oxygen for a water-bottomed mile.

When I was ten, the river reached for us again. On the way back from church in Brunswick one Sunday afternoon, we stopped at Everett, a windblown village near the river delta, to visit folks my father had met. They were docked permanently on a houseboat and my father, a part of whom yearned for a more bohemian and simple existence, was curious about their life.

While the adults talked, the three of us younger

children wandered to the open end of the boat. A railing welded of galvanized pipe circled the deck. We'd been told of course to be careful, and we were, since none of us could swim. We sat quietly, intrigued with the fat, foreign river and its line of docked houseboats, listening to the gurgle of current against wood. The river seemed impossibly huge, deep and mysterious, as if it had a mind and thus a purpose. My two brothers, Dell and Steve, sat side by side in an aluminum lawn chair next to the railing; I sat in another, nearer the wall.

"How many thousand gallons of water do you think pass here every minute?" Dell asked.

"Millions," I said.

"Billions," said Steve.

Dell was peering off at the river. "Is that an alligator?" he asked suddenly, and stood up. When he did, the lightweight chair tipped outward. Steve was closest to the water and in a second was overboard.

I heard my father running before my mouth loosed a syllable. He was at the far end of the fifty-foot boat when the water swallowed hugely, hissing against itself, and he knew immediately the sound of loss. He sailed over the metal railing like a high jumper, in his best Sunday suit, white shirt, black pants, cuff links. There wasn't time to kick off his black polished shoes. I remembered later that he'd loosened his tie in the car outside Brunswick and draped his Sunday coat over the seat. His dive was a masterful feat driven by fear and sheer unconditional love; Steve's black hair, blown and wisping in the current, sped toward the next houseboat thirty feet away. Any minute this youngest son would disappear beneath it and certainly drown.

Swimming in an adrenaline river, my father reached him just in time, tackled the invisible body of his son, and stopped them both on a wooden beam on which the next boat floated. With ropey forearms Daddy lifted Steve, then climbed up, streaming. For a few moments no one moved or spoke. Daddy wouldn't let go his hold but sat swallowing river water, chest heaving, looking back at us with wild gratitude in his eyes.

After that he determined we'd learn to swim. It was hard to find a public yet private place where no one would see us skimpily dressed. Near the river he discovered a half-acre secluded lake with a dirt beach, cloaked by young maples and other brushy trees—actually a borrow pit from road building, I think, where people rarely came. If anyone turned off the highway and drove down the rutted road toward us, we were instructed to immediately leave the water and jump into the car, wet or not, even if we had only just arrived. Spirits lighten in water, and the family would begin to play, splashing each other and laughing. Daddy would flirt with Mama and tease her, and she would laugh.

We used flotation devices homemade from two empty milk jugs tied to each end of a two-foot length of rope, worn across the chest beneath the arms. The jugs flowered from the armpits like bulbous wings.

My mother had been thrown into a swimming hole, a deepened place on Little Ten Mile Creek, when she was a girl. Unable to swim, she had nearly drowned before her brothers managed to fish her out. She had a long fear of water that, married to my father's long fear, solidified in us. I learned to dog-paddle but not without milk jugs.

We were loved. We were desperately loved.

I have witnessed my father willing to lay down his life for us, like the time we were pouring cement for the foundation of the new house. Daddy sent Steve to switch on the tumbler of the concrete mixer. Next time he turned around, Steve was paralyzed at the mixer, both hands clamped on the steering bar. The mixer had shorted and 220 volts of current poured through its metal into his young son's body. Instantly—quicker than instantly—Daddy was flying toward Steve, leaping at him, snatching him from the grasp of electricity.

He could not have survived losing one of us, he knew, so he was afraid. Irrationally afraid. He guarded us like a warden—exactly like a warden—and he taught us always to think of safety first and to be prepared for the worst and not to trust and always to be afraid. If he and Mama left the four of us alone while they went to look at a junker someone wanted to sell, Daddy left a litany of instructions still pressed in my brain.

"Keep the door locked. Don't let strangers in the house," he'd say. We were not to answer the door, not for anybody. If we must speak to someone at the door, speak through the window. To disobey would mean instant and harsh punishment, a whipping with the belt.

Always there were guns. We four children knew that any one of the weapons lying around might be loaded and we were never to play with them. We lived just outside of town, but neighbors weren't close enough for us to run for help. Any time Daddy left the house, he reviewed the arsenal.

"A rifle's leaning behind the door," he'd say. "It's on safety and there's a shell in the chamber. You slide this, and it's ready to fire. Here's my .38 under the cushion of the chair. It's semiautomatic, so you have to push safety off on the side and pull the trigger."

He was obsessed with protection, maybe because he'd had to protect himself for so long. On Sunday afternoons we'd target shoot. We'd go to the farm and set up a line of drink cans or a pasteboard box drawn with a spiral of circles. Before I was ten, I was shooting a .22-caliber pistol my stick arms could barely support.

I was deathly afraid of guns. I knew their power. The last thing I wanted was to kill somebody. I thought that as soon as Daddy handed one of us the .22, the gun of its own volition might turn on us and fire. He watched like an eagle as we shot, distractedly wiping sweat rolling from his temples, making sure we kept the gun trained on the woods. My eyes wouldn't leave the pistol either, not until it had made its round, and counting, I knew its six chambers were empty. Even then I'd never pass in front of a barrel.

Marksmanship was as high an achievement as straight As, and target practice was competitive: whoever knocked the can off the post or put a hole in the bull's-eye of the wobbly circle was hero for the day. One round Kay's shot would fall closest to bull's-eye, the next round Daddy led again. We took shredded targets home to brag over. We all got to be fine shots—the eyes, the arm's endurance, the responsibility—three holes in an area the size of a fist, a plum, a quarter.

Daddy was not a man either quick or lavish with praise, but when pleased, he could not hide his pleasure. His approval was a thin fertilizer and never enough, but he loved us fiercely, we knew, or we wouldn't be practicing at targets at all. Hard love.

"Okay, it's Annie Oakley's turn," was all he'd say. "Try to hit the box at least." Later we might overhear him bragging about us to an uncle, and then we would know we had done well. Only then.

His scorn, on the other hand, would wither a stalk of corn, and for fear of wrath, we forced the best of ourselves. We were not half-assed. We swallowed fear and pushed back tears and willed the gun to hold steady and the bullet to knock the can from the far post.

Neither was mine a part-time Daddy. He was never absent, every day present. Any work that consumed him also consumed us. Except when he was ill, he never left us.

Most of my memories of the Altamaha come from a rafting trip taken with him and brother Steve the summer I was eighteen.

Until 1870, the Altamaha River was the only mode of transportation for settlers of Appling County. Pole boats and rafts, loaded with corn, cotton, sheep's wool, skins, tallow, and other raw materials, floated down to Darien, where the backcountry goods were shipped to Savannah and other ports. To return, the raftsmen poled upstream against a strong current, bearing home salt and other supplies. Longleaf and cypress logs, squared into two- and four-foot timbers and lashed together, were floated downriver when timbering was young. Rafting the Altamaha, then, was a natural thing to want to do.

That winter I'd read *Walden* at the urging of a clandestine boyfriend who came to visit me in the stacks at the public library, where I worked. The book had made me think, made me get down on my knees in front of an anthill in the junkyard to look at its activity, and had

made me simplify my room—did I own the throw rugs
or did they own me?

One Sunday soon after high-school graduation, Daddy
walked into the kitchen where Mama and I cooked dinner—my sister long gone, graduating from college—and
announced out of the blue, "I think it's time we went on
our river cruise." My father had lots of schemes. He always
wanted to camp our way to Mexico and see the West, and
we got as far as rebuilding a pop-up, pull-behind camper.
It had tin sides that went straight up into an A-roof and
looked more like an outhouse on wheels than a camper.
We would go sit inside and plan where we'd all sleep, how
we'd keep dishes and jugs from sliding off counters by
nailing in retainers, what we would carry. My father loved
to dream, and hoping hard, we dreamed with him. He
wanted to show us the Grand Canyon. Many winter nights
we sat in front of the fire, shelling pecans and eating
them, talking about all the places to see between here
and Mexico.

"I'd like to see the desert," Daddy would say.
"Roadrunners live in the desert."

Dell would know how fast they run, and pretty soon
somebody would bring down an atlas, and I'd be figuring how many miles we'd travel and multiplying out how
much we'd spend in gasoline between Georgia and
Arizona.

"That's if we get twenty miles to the gallon pulling the
camper," Daddy reminded. We were not eager to go in
the camper. It was unbearably ugly, even to us children.
By the time we'd loaded up, we'd look like the Goat Man,
we felt sure.

The Goat Man was an old gentleman who wandered

perennially through the South in a carriage pulled by a small herd of goats: wanderlust and poverty wrung together. His wagon was tacked with hubcaps and all manner of junk he'd found by the wayside, and he was thin and dirty with a wispy gray beard that covered his shirt pockets. He had extra goats tied behind and some of them he milked. Living on U.S. 1, we saw a good deal of interesting traffic, including Yankees with New York and New Jersey license plates who stared at us without expression. But the Goat Man was indeed a sight to behold, and if anyone saw him coming round the corner, she went yelling for the others and we lined up along the highway to have a good look. The Goat Man never stopped, spoke, or looked at us, and seemed oblivious to the cars that lined up impatiently behind him, waiting for a chance to pass and gawk.

We never made it out West.

For years Daddy had talked about rafting the Altamaha from start to sea, as the early raftsmen had done, on a boat without a motor. Our neighbor across the street, Green Deen, had captained the last steamship to operate on the Altamaha. He loved the river, and Daddy would talk to him of his idea. Mr. Deen thought it a swell one.

"Shore," he'd say, and they'd talk about landings and places to camp: Falling Rocks, Jack Sutt Landing, Hell's Shoals. So when Daddy entered the kitchen and announced, "It's summertime. You'll be leaving for college soon. Now's the ideal time to raft the river," I wasn't surprised, but I decided not to let it end at dreaming.

"When?" I asked.

"Maybe next weekend," he said.

"Let's do it."

A few times during the week he hedged in his commitment, but I threw myself into preparations and pressed him to follow through. And it happened. We decided Steve and I would go, and Dell, who worked Saturdays weighing nails and mixing paint at Lewis's Hardware anyway, would remain with Mama.

Early Saturday morning we drove to the Lumber City landing and unloaded the heavy, sixteen-foot, flat-bottomed, motorless skiff into the Ocmulgee River. The outdoors was all so new to me. I'd never spent even one night in the open. I was wearing pants for the first time in my life, a pair I'd found at the dump.

From beneath the old railroad bridge we waved good-bye to Mama; she was to expect us back the following day.

"We'll stop at Deen's Landing and call you," Daddy yelled.

"OK."

"Don't get worried until late tomorrow."

"Be careful."

We were barely in the channel when Steve noticed water pouring from two half-inch holes near the top of the overloaded stern.

"Daddy! It's leaking!"

Daddy leapt to the stern. "Get this gear to the bow," he ordered. "Where are those extra corks? Get 'em for me."

Extra corks? To fish with? To plug holes? What mishap had he not prepared for? In seconds he had the holes plugged.

We knew nothing about keels or currents. Instead of staying in the channel with its bow addressing the sea, our boat spun, sailed sideways, went backward, and careened from the "white side," which is what raftsmen

called the north side of the river, to the "Injun side." Only
by paddling could you right it, and paddling the barge
was exerting, since it rode over two feet out of the water.
The Ocmulgee was dark with overhanging vegetation,
full of curves, and the current took us straight for each
of them.

Part of the aging process of a river is continually to
deepen its curves until two bends connect, breaking
through land and leaving an oxbow lake behind. On outside bends the river gnaws at trees, which cling to shifting
banks, half their roots reaching back into the bluff, the
other half smitten by water. When their hold weakens,
they tumble and become snags.

No amount of frantic backstroking could keep us
from tearing into the downed trees if we drifted too close
to the curves, so we had to anticipate curves and paddle
out of them.

"You don't just sail along like I thought," I said.

"When you snooze, you lose," Daddy said.

In the spirit of Thoreau, since I had his *Journals*
along, I wrote an essay about that trip. "Bends are the
meat, and snags the poison," I wrote. The metaphor must
have meant that bends pull you in and then snare you
with snags. Closing the essay, I said, "Respect the river for
its power and silence, and for its grip on you, but don't be
afraid of it. Feel your charge and its charge as opposites
attracting." Reading that now embarrasses me; I was so
naive. We are no match for nature; nature is of God, of
eternity: deadly.

At seven-thirty, Daddy landed on a sandbar long
enough to fill the bottom of a stainless-steel foot tub with
sand and set it in the stern. Steve and I collapsed into

giggles watching him blow and rearrange and blow, finally igniting a heap of charcoal and kindling. We wrapped sausages and potatoes in aluminum foil and finally the fire was blazing well enough to start them baking. I mixed mustard and molasses in a can of beans, balanced it in the fire tub. At this rate it might be hours before we ate. I was starving.

It was dark by the time the potatoes were soft, and we ate by kerosene lantern light. We drifted along, savoring the moments. Before we unwrapped slices of chocolate cake, we realized the river was thrashing about, fretting, and pushing us toward the inner bank. It was also widening. Daddy dropped his cake and grabbed a paddle, to reenter the current.

"It's the Forks," he said. The Forks was the confluence of the rivers, site of Creek encampments, crossing of major trade routes, both overland and riparian. I believe the Lower Creek came here when the shad were spawning. Supposedly they used the shad on the Altamaha for fertilizer. The native fish spawned upriver then died, and from March 15 to April 15, dead shad drifted back down the river. People remember seeing thousands of them drifting down as late as the 1940s.

The Oconee, a stronger river, pushed the Ocmulgee away, exiling us toward the bank. Daddy, fighting the water with his old paddle, did little to contradict the water's force. Instead, we hugged the edge until we passed the fighting waters and entered a new, more peaceful mainstream.

From the tumultuous marriage of the rivers, the calm Altamaha was born. The full moon, hidden by trees on the Ocmulgee, hung overhead when we turned into it.

"We picked a good night, didn't we?" I said softly.

"No, *we* didn't pick it," Daddy replied, just as softly. He was unfailingly devout, always grateful.

That night we put down a makeshift iron anchor and slept in the seats and in the prow. Daddy thought it safe enough to doff the life vests, which we had been wearing the entire time, and it was a good thing, since they were awkward and unwieldy. In the long night, few mosquitoes traversed the broad water to annoy us, and once I awoke terrified, not knowing where I was or why my bed rocked. Then it all came back. I looked around. We were not where we had anchored, but somewhere downstream of that, which meant the anchor had not caught and we were drifting slowly, dragging it. It didn't seem to matter. Until I slept again, I surveyed the night. The river, lit silver by moonlight, was eerie and spooky, a world of its own. From the swamp came odd birdcalls, great horned owls and yellow-crowned night herons.

At dawn, sliding silently through thick fog, we passed McMatt Falls. One by one we wakened as if from a long dream. This was how Rip Van Winkle must have felt that first morning of his renewal. Daddy cooked grits, fried eggs and bacon, even managed to toast bread by laying it across a rack on his foot-tub fire. Later, when the morning sun still shone softly, we landed on a white sandbar to wash up, brush our teeth, change clothes. In thigh-deep water, Steve's billfold fell from his pants' pocket.

"It was right here," he said, planting his feet deeper into the sandy riverbed and bending at the waist to feel bottom. "Come help."

I waded to him and began combing underwater with open hands, scribing circles in the water with my feet.

"Son, you can kiss that billfold good-bye," Daddy said. "It's a quarter mile from here by now. What did you have in it?"

"Two or three dollars," Steve said. "I can't remember exactly. Some pictures. My learner's permit." He was fifteen and able to drive if accompanied by an adult.

"Nothing you can't replace," Daddy said.

All day we passed people fishing.

"Neighbor, you having any luck?" Daddy would call.

"No, not too much today. Caught a lot yesterday."

"What you fishing with?"

"I got a little bit of it all."

"Say, how far would you say it is back to the Lumber City bridge?" Already we were pulling away.

"Oh, I don't rightly know, but I'd say it's twenty miles," the fishermen would call as we drifted out of earshot.

How far to the Uvalda bridge? How far are we from where the rivers fork? How fast does the river run? If you're drifting, how many miles an hour are you going?

One fisher would say, "Oh, you have about three more miles to go. You'll make it by six o'clock." The next would say, "Good gracious, I'd say it's fifteen miles down there. Long way."

More than once we barely avoided crashing into snags and had to clamber out of the limb claws of downed trees, and once, midafternoon, we bottomed out on a sandbar. I rolled overboard to push us out of the scraping sand, when suddenly the amber-colored water went black and the bar dropped away. The skiff rushed out of my hands and left me spluttering in the suddenly deep water, the life vest buoying me. This was more excitement, but I swam badly, gratefully caught the paddle Daddy extended and heaved myself into the boat.

Late that afternoon, red as fireballs and exhausted, we pulled into Deen's Landing to call Mama. She met us at the ramp under the highway bridge, just upstream of the nuclear plant.

Pine Savanna

Within the longleaf pine ecosystem lies a smaller, more unique community, the savanna, a broad plain of herb bog that joins the longleaf pine flatwoods to the swamps. Sometimes this zone is no more than a few yards wide and sometimes it is acres of soggy and supple prairie. These are called seepage bogs because water inches across them, draining the high ground, filling sloughs and streams and pools downslope. Unlike tree-dominated swamps and forests, savannas are composed of herbaceous vegetation and open to full sunlight. Here and there a lone pine rises, although most savannas are too wet to support many trees but dry enough for occasional summer fires to pass through. Wiregrass and other unusual grasses join a suite of bog plants exotic and bizarre in their adaptations to these conditions.

These seepage bogs are among the most botanically diverse on the planet. More than fifty species of flora have been counted in one square meter of savanna. Three of the world's four carnivorous plant families exist here: six pitcher plants; twelve bladderworts and six butterworts; and five sundews. Nowhere else on the continent can this diversity of carnivorous plants be found. Such is the variety and pageantry that these have been called "pitcher plant bogs."

Thousands of trumpet pitcher plants flood the savannas, their knee-high hollow leaves covered by overarching hoods. Some like bonnets, some like parasols, the hoods shield the long green tubes of leaves, which, lathered with nectar and fashioned with hairs, lure insects inside with flowerlike scents and bright colors. The bottoms of these pitfalls is a simmer of digestive enzymes that distill from the insect carnage minerals the plants cannot obtain from the acidic soil. In April and May, trumpet flowers are delicate yellow hands rising from wet ground. Almost exclusively, bumblebees pollinate them.

Parrot pitcher plants look like the beaks of parrots, reclining toward the ground. Their flowers are dwarfish, garnet, and of all the pitcher plants, they love the wettest ground. Frill-edged white-topped pitcher plants, which when first discovered were misidentified as a cousin of jack-in-the-pulpit, stand alongside their two-inch burgundy flowers. Hooded pitcher plants are probably most common.

The interior of the pitcher leaves is a unique microenvironment, a rudimentary living cave supplied with trapped prey. A number of species have evolved to inhabit pitcher plants without being eaten themselves. The immature stages of one mosquito occur in water caught on purple pitcher plant leaves. The larvae of certain flies inhabit *Sarracenia* leaves, where they feed on insect bodies. A moth of the genus *Exyra* spends its larval stage inside pitcher plant tubes, feeding on leaf tissue.

Deeper in the grassy vegetation, the butterworts

bloom near the ground. They are carnivorous plants with a roseate of seductive basal leaves that are buttery, sticky slopes unwary insects plunge down. They bloom blue, soft yellow, rose, and white. One of them is red-leaved.

Next to the earth, red-bristled sundews, tiny but deadly, offer up their tempting honey. They look like bouquets of tiny, exquisitely bristled flyswatters. The entire plant, twenty or so of these flyswatter leaves linked basally, is often no bigger than a cat's ear and just as soft. When they detect motion, the miniscule and glistening palms of sundews slowly close around struggling prey.

Bladderworts grow in water, smothering shallow ponds and ditches with lavender or gold blooms. They trap aquatic microorganisms and minute insects in translucent pinhead-sized bladders—really modified leaves—underwater.

Savannas are magnificent wildflower gardens. Something is always blooming: grass pink orchids, rose pogonia, rosebud orchids, ladies' tresses. In the heat of summer the fringed orchids are torches through the meadows. Blue-eyed and yellow-eyed grass, white-eyed sedge. Meadow beauties. Fall brings on the composites, purple spires of liatris, also called blazing star, brown rayless sunflowers, goldenrod, bigelowia, and coreopsis. The aster *Balduina* pools like orange juice in the wettest places and clumps of pearl-tipped hatpins stick the carpet of forbs to the flat earth. White violets hover low.

These grasses: *Aristida* (wiregrass), toothache grass, *Andropogon*, dove grass.

Longleaf pines are important to neighboring savannas mostly because of fire, which historically crept out of

the burning flatwoods and moved through the pine savannas, all the way to the swamp. Fire repressed the advance of woody shrubs, which are wont to move out of
swamps and colonize savannas, and created a blank slate
upon which plants flourished anew. Like longleaf forests,
pine savannas became fire-dependent. The absence of
fire results in the elimination of bog species.

I've been to savannas many times and have even slept
in them. At evening in a Southern coastal-plain savanna,
chuck-will's-widows call from the piney flatwoods a quarter mile upland. Not a rustle or hoof fall punctures the
peace, except one moaning howl a mile away that might
be coyote or wildcat. Half the moon, like a broken dinner
plate, poises directly overhead and the stars begin to appear, first in the nimbus of moon, then one by one across
the lit sky. Soon a canopy of purple velvet is sewn with diamond flecks above the thick and hillocked grasses.

The dreaming is deep.

In the morning, a strip of pink pools through the slash
pines in the lowland east. The fog is a garden wall made
both of stone and imagination. In the garden a million
million spiderwebs are spun of strands of dew—dream
catchers, wind nets, hammocks of dawn. There are thousands of them, a revolution of spiderwebs in an anarchy
of fog. It is like an ocean of webs, every tussock slung with
a diaphanous nightcap.

They melt into the rising sun at the rate the stars
formed—so little of nature happens in revelation, but
rather in mad warps of time. A pygmy rattler curls on a
warming pad of grass, high off the ground, soaking in a
suffusion of heat. In the crayfish burrows two-toed am-
phiumas and mud salamanders live their silent lives. A

pair of bluebirds, awesomely blue, dawdle in a gnarled pine. Across the wide savanna, fifty to a hundred tree swallows dip and bank, snapping up insects only they can see. Meadowlarks and red-winged blackbirds sing to the bird's-foot violets and to the odd flightless grasshoppers and spittlebugs who can live no other place. In the bog pools, the flailing larvae of endangered pine barrens tree frogs stir microscopic clouds of mud.

Driving and Singing

Sometimes in warm weather and even in cold, to escape the house or the endless work, we would go sit in the junk cars with the windows rolled down, and we would pretend to travel to far-off places.

"Where do we want to go today?" I'd ask Dell.

"I've been wanting to see Canada," he'd say.

"That's a long way from here," I warned him. "I don't know if we can make it back by supper."

"I bet we can," he'd say. "Just drive fast."

I pushed in the useless clutch and threw the transmission into first. I began to make the ummm-ummm noises of a motor accelerating and went through all the gears, whipping the steering wheel this way and that, except it didn't move far since the flattened tires of this particular car were partially buried in soil.

"My goodness, we're going sixty-five already," I'd say.

"You'd better watch out for police."

"I can outrun police."

"Look at this. We're in the mountains. This is a big one we're climbing. Wow. What a sight. Oh no. We're going downhill now, and I can't get the brakes to work."

I'd be pumping furiously. "Help me. Pull the emergency brake. Help!" Dell would make a general uproar, flinging his arms here and there, grabbing at levers and knobs.

"Uh-oh," he'd say. "Look in the rearview mirror. You got stopped already."

I'd glance up at the mirror, but it was too moldy and dusty to see through. So was the back windshield. I'd stop and talk to the invisible officer through the window, either being very rude or very solicitous, and soon, thirty-dollar ticket in hand and a promise to slow down, we'd be on our way again.

"Fartbag," I'd say then, and Dell would agree.

"Hogbreath."

"We weren't going eighty. No way."

We drove past the White House, through the Amish country of Pennsylvania, through the congestion of New York, around the Statue of Liberty, crossed the Great Lakes on a ferry, and sailed through Canada. There was snow all over the place, and we got out of the car to take a look. Dell spotted a white wolf, which prompted me to see a polar bear and then a snowy owl. Then we changed drivers, turned around and drove home.

Never moving an inch we saw many faraway places we knew only from *National Geographic,* one of the few magazines allowed in the house. That was how we knew about snow, from pictures we'd seen of the South Pole or from Christmas scenes, with bundled Santa Clauses sleighing across snow-covered roofs and all the world white. When it finally snowed, we could see and feel what snow was like.

It was January of 1977, the winter I turned fifteen, when it happened. The winter had been extraordinarily cold and harsh. On January 20, school had been canceled for the inauguration of Jimmy Carter, and for the first time in my history, a few days later the schools were closed for weather. This January morning Mama had gotten us up, as usual, then flipped on the kitchen radio. The announcer was excited. "It's snowing!" he said. "No school today." Mama called up the stairs, "Look outside! There's snow." She had seen it once before in her life. It was barely light enough to see outside, the sun just rising, but Mama's news was truth. Puffballs of snow fell from the dark tumbler of sky. My window ledge was mounded with fluffy crystals, and so pretty.

Mama made us eat breakfast before we bundled up to go out. I wore two pairs of socks and a muffler around my head, all I could find, since even Mama—our Georgia Mama—wasn't prepared for snowstorms. Outside, the roof was frosted white and the limbs of the trees mounded with snow. Smoke wisped from the chimney. For the first times in our lives we made snow angels, falling prone and sweeping wide our arms and legs as Laura Ingalls had done. We hesitantly packed snowballs, having never felt this odd manifestation of winter and not knowing what to do with it except try what we had read. We built a snow person. When our fingers and toes were nearly frostbitten, we came inside and warmed by the fire.

Later in the afternoon, when ice on the roads had melted and the snow was disappearing fast, we piled into the car and rode around to see our world suddenly different and to see how others had responded to a day of snow and no school. All through town children were building snow people and running and playing in the cold surprise.

In reality, riding around in the car was one of the major sources of entertainment for the family. I have spent hours and hours—only one or two at a time but on uncountable occasions—riding around the countryside. After supper Daddy might suggest "Let's go riding." Riding is what we would be doing any Sunday afternoon. We all piled into the backseat of the Mustang we drove or in the back of a truck if one was running, and Daddy drove slowly, twenty-five or thirty miles an hour. Often we rode out toward the farm, the seventy-acre woodlot out Old Surrency Road Daddy had managed to buy, and sometimes we stopped by Granny Ray's for a short visit or rode out to Grandmama's to see how she fared. If there was something interesting—a wreck, a chain gang cutting roadside brush, a hitchhiker—we craned our necks to see. If travelers had broken down, we stopped to help them, and if people we knew were outside, Daddy would brake to speak. Teenagers kept the strip from Dairy Queen on the north side of town to Hardee's on the south (our two fast-food restaurants) cooking, and we rode through Baxley a couple of times to see what was happening, although mostly we drove slowly along the dirt roads. Daddy knew the county inside and out. If he didn't know where a road came out, he would follow it, and we would drive until we again reached a crossroads or a farm we recognized. Before heading home, we stopped for bread if Mama needed a loaf or for milk and cornflakes for Sunday supper.

We rode to see what could be seen. We rode because there was nothing else to do. We rode because it kept us together, close in the car where we could talk. If we got crowded and irritable, Daddy made us be quiet, then he asked us quiz questions.

"Okay, who can answer this? If thirty-three and a third is a third of a dollar, what's a third of a dollar and a half?"

The answer would take a bit of dividing.

"Fifty cents!" Kay would yell.

What we liked best was riding in the back of the truck at night, lukewarm stars overhead and mosquitoes foiled by our motion. Before us in the darkness the wild green eyes of animals would flare before they turned and went out. Ghost deer paused at the edge of wilderness.

Sometimes as we rode we sang songs we knew, some gospel, some not. We belted out "The Old Rugged Cross" and "When the Roll Is Called Up Yonder" and "Away Down South in Dixie." People have said that they could hear us coming up the roads, driving and singing.

The Kindest Cut

There is a way to have your cake and eat it too; a way to log yet preserve a forest. Leon Neel knows how.

He is an ecological forester in the pine woods of southern Georgia and northern Florida. A protégé of ecologist Herbert Stoddard (colleague of Aldo Leopold who pioneered the recurrence of fire in managing long-leaf forests), Leon applies ecology in such a way as to preserve a forest intact while extracting economic benefits. Known as single-tree selection or uneven-aged management, Leon's silviculture selects by hand individual trees to be harvested and leaves multigenerational or multi-species growth in a handsome, functioning grove. It is an innovative alternative to clear-cutting, proving endangered species can exist in a working landscape.

"You never terminate the forest; you terminate individual trees," Leon says. "You never regenerate the forest; you regenerate individual trees."

"I think the whole forestry profession has got to re-organize," Leon says. "Now it is dominated by big industry, which is concerned only with making money." He proposes a system that recognizes, in addition to timber production, the value of aesthetics, abundant wildlife, and a rich diversity of plants.

Leon takes me to see his prize grove. We stand in the company of tall, sprightly longleaf pines, their crowns a nebula above us. Early morning sunlight trickles in soft, diffused shafts across the dew-soaked vegetation, in which orange-fringed orchids and purple, fluttering meadow beauty bloom. A quail calls, "Bob WHITE, Bob WHITE." A red-bellied woodpecker dips by, and we wade out to a healthy clump of *Sarracenia minor* almost hidden among the wiregrass and bracken fern.

It is the most elegant forest I've ever seen.

Leon explains that for forty-five years no trees have been taken from these five hundred acres, part of an extensive plantation owned by a wealthy industrialist. Many of the pines around us are over two hundred years old and have outlasted any human on the face of the planet.

"I use this forest as my model," Leon says, then pauses. "This is what heaven should be."

Leon is wearing khaki work pants and calf-high work boots of dark leather, no cap. His silver hair matches the rims of his glasses.

"By 2050 there'll be twelve billion people on the earth demanding food, water, TV sets, whatever people demand," he says. "I really am worried about it." He wonders what the population surge and the demand for goods will mean for the thousands of acres of woodland he has cared for, in the true definition of the phrase, for over forty-five years.

Winding farther into the woods, we pause to look at runner oak and low-bush blueberry, both important provender of quail. Pond and slash pine grow in lower pockets in the longleaf. A couple of Bachman's sparrows rise out of a stump fringed with palmettos, and an eastern kingbird, its slate tail hemmed with white lace, perches on a twig. In the distance a mourning dove coos its haunting song.

"The character of these woods changes from minute to minute, day to day, weather to weather," Leon muses aloud. "They're beautiful under any conditions, but their character changes." It pleases me to hear woods spoken of in such tender and perceptive terms.

Leon's woods are full of deer. Quail and turkeys abound, and gopher tortoises. Fox squirrels shimmy up the flaky, gray-brown boles of the pines, and hundreds of migratory songbirds stockpile provisions. There are great horned owls and red-tailed hawks. Red-cockaded woodpecker candle trees pepper the woods. Leon, through his work, is responsible for the largest population of red-cockaded woodpeckers left on private land anywhere—the Red Hills around Thomasville, where large, adjacent, quail-hunting plantations spared the pineywoods that red-cockadeds need.

"You've got to think beyond your own life if you want to perpetuate the red-cockaded woodpecker," he said.

The piney flatwoods he manages are burned regularly, often annually, in order to suppress hardwood growth, rejuvenate the soil, promote diverse herbaceous growth, stimulate flowering and seeding, and create wildlife habitat.

Although cutting trees is the hardest part of Leon's job, he realizes most landowners can't afford to immortalize a forest for the sake of the forest and its creatures. Property taxes come due and the land has to support itself. In Leon's system, trees are harvested every seven to ten years. On any one cut he takes no more than 20 percent of the volume. Instead of returning a large sum of money to a landowner at one time, Leon's forestland yields smaller amounts over a long period.

In the selection of trees for cutting, he uses a number of criteria. A tree with blisters, defects, or knots is sacrificed first, along with an off-site species, meaning a kind of tree that did not historically grow there but has been planted or otherwise introduced. Sometimes a tree is cut in order to release (or open up space and sunlight for) another. A tree that harbors wildlife is spared, as are older, straighter, healthier trees. Trees in areas that harbor rare or endangered species are passed over.

We have been winding through a forest road as we talk and have passed onto an adjacent piece of property, one currently being logged. The crews would be here now if it weren't for recent heavy rains and Leon's insistence on waiting until the ground dries. When the ground is this wet, heavy machinery wrecks vegetation.

"But listen," he says, "We have excellent loggers who recognize the value of this system. They want to protect the woods as much as we do." He pauses. "We also have scoundrels, the ones who cut out and get out."

"We get paid by what we cut," Leon continues. "But I have refused to cut trees time and time again." Then he adds quietly, "If I wanted to be rich, I could be rich."

Leaving

When I was eighteen and away from my town, I dived
recklessly and surely into the world, not because it was a
form of rebellion, as people might think, but as a form of
healing and survival. Mama and Daddy wanted me to be
a schoolteacher, a job that guaranteed an income and
the best employment Appling County offered, but I had
my heart set on literature. Faulkner and Styron and
Warren, I felt, had saved me; I wanted a life constructed
of books.

"You can write in the summers," Daddy said.

I chose to attend North Georgia College, a small military school in Dahlonega, a mountain town that had suffered a gold rush in the mid-1800s, producing enough
ore to gild the dome of the state capitol and make a
semitourist attraction of the place. The town is located
on the edge of the Chattahoochee National Forest,
foothills of the Appalachians, where fall torched the
mountainsides to gold, orange, and red; real snow fell;
and clear streams sloshed across their rocky beds and
emptied into fast, boulder-strewn rivers: the Chatooga,
the Jacks, the Nantahala.

Dahlonega was built around a square, where the court-house stands, with old wooden storefronts surrounding it—a potter's shop, an ice-cream restaurant, a real-estate office, a health-food store. The town was supported by the college and by fall tourists who came to see the leaves.

I chose North Georgia because my sister had gone there and because scholarships paid my way. I was there two years, and it was a stepping-stone into myself.

I lived in a red-brick, ivy-covered dorm in a pinched room with metal furniture. My roommate, a self-confident red-head from Atlanta, spent more time with her boyfriend than studying. Men were not allowed on the halls, and many nights she did not come home. When she did, she slept in. I was up early, by six, when reveille sounded and the thud of marching feet marked time to the platoons' chants. *We're the men from Alpha Corps. We don't like the girls no more.* Before dawn, the senior officers were out, running the frogs, or freshmen. (Military was optional for women, most of whom opted out, but mandatory for men.)

Except for the brutal hazing and the grueling demands placed on frogs, I felt like one of them—tortured by my lack of experience and my newness. So many things I had never done, and now I was on my own I could choose for myself, an enormous responsibility. I had not dated, nor experienced much of a social life. I had not played sports. I had not been to a movie or to a bar or dancing.

Financially, if not emotionally, I was independent of my parents, who had no money for college tuition. With savings from my after-school job, and with the three dollars an hour I earned from work-study in the chow hall, I would manage. I was on my own.

The first friends I made were three young men who also worked in the cafeteria. They had chosen North Georgia not because of the military but because it was close to home, Atlanta, and near the wilderness. Our friendship had nothing to do with romance. Friendship with men was familiar territory, since I had brothers, and they treated me like a guy. I got drunk for the first time with them. It was a Saturday night in October, and we had gone to Gainesville, thirty minutes away, the four of us, to see a movie. We were drinking Boone's Farm wine, and once the alcohol hit my bloodstream, I was uncontrollably giddy.

On the ride home we stopped for hamburgers at McDonald's (Dahlonega was too small for a fast-food restaurant at the time), and I remember standing at the counter so dizzy and drunk I couldn't read the menu board. I found it superbly funny. At that point my friends did not know what to do with me, except take me home, and I did not know what to do with myself, except go.

Another time we all went camping on the Appalachian Trail. I carried a cheap cotton sleeping bag that had come in a load of junk, and I was ill prepared for night temperatures in the mountains. I was so ignorant. We built a fire, and I lay as close to it as I could and still nearly froze.

A military school is full of rules. Men wear uniforms except after certain hours. Frogs wear them all the time. No public display of affection is allowed. Uniformed cadets must salute outranking officers. When the bugle sounds retreat, everyone must pause until the cannon fires and "colors" is played. Cadets must salute the flag.

At midnight on weekdays and 2 A.M. on weekends, the dormitory doors were locked, whether you were in or out.

Wanting to be more like other girls, I tried cheerleading for an intramural team. I went to Mardi Gras with the band, as drumstick girl, which meant I picked up the sticks the drummers dropped. I inherited many of its members as friends from my sister, whose boyfriend played the tuba. They dressed me in a brown uniform like theirs. Somehow, however, I always wound up with the wild crowd, the liquor drinkers and risk takers. It was what I had to do.

One night after dark I went with a carload of students to night-rappel. We drove up into the national forest where a mountain had been blasted away for the highway, leaving a one hundred-foot cliff that looked high and eerie in the car lights. With flashlights we unloaded gear and put on extra coats and climbed a faint, sloping trail that led up the backside of the mountain, to the top.

One friend knew what he was doing, although he was the wildest among us, the one whose idea this had been, and the one who owned the gear. He checked the rope and snaps and knots. He knew the trail and the tree to tie to.

I did not go off the cliff first. When it was my turn, Hal strapped me to the rope and demonstrated how to brake with a wrist-turn on the rope. I backed to the edge and looked down into darkness. I heard voices and saw the bobbing of a flashlight, seeking out the trail back up. The scene was a metaphor for my arrival into the world, the real world—me standing at cliff's edge, unable even to imagine the route down.

"Heads up," Hal shouted, and I eased into the void.

The rope wanted to sing me down, but I braked breathlessly, bouncing into the cliff wall, where shrubs and long grasses had taken root. My knee hit a protruding rock. I let off the brake, but slowly, burning the glove leather to avert crashing against the unseen ground. The danger, the excitement, the darkness, the thrill, the passing car lights that caught us hanging—this was what I had been missing.

The same friend persuaded me to skydive too. He wanted to be my lover, but I was too afraid of him. He was the kind to self-destruct. Still, he was fearless, and no matter what it took, I had to learn fearlessness. The skydiving school was close to Atlanta; a carload of us drove there one Saturday morning. All day we were instructed on how to jump, how to land, what to do if the parachute didn't open. We signed waiver forms. Then, late afternoon, a plane took us up.

One by one we had to climb out the open door onto an airplane strut. We were jumping static line, which meant the chute got opened for you. All you hoped was that it had been packed correctly. You fingered the rip cord of the reserve.

The ghost of a cadet was said to roam the men's hall at college. The cadet had been a skydiver so interested in the sport that he purchased his own chute. Not liking how the chute was made, he remodeled it so that the reserve was on his chest, not his belly. One day his main chute malfunctioned, and when he pulled the reserve, which he had sewn by hand, the whole thing ripped away.

I crouched in the belly of the plane, watching the ground far, far below. I might not make it back alive to that good earth. Yet when my turn came, I crawled to the door and eased out onto the strut, facing forward. Even though the plane was idling through the sky, the wind hit me so hard it flung spittle from my mouth. It pushed my cheeks this way and that. Even then I could have changed my mind, but I had to learn courage—I had to—even if it meant death. If I could do this, I could do anything.

I looked back at the instructor. He looked at me. I nodded. He made a motion and smiled, and I let go, tumbling out and away from the plane. For a few seconds I was flipping slow-motion through the heavens, then I heard a loud snap, and with an enormous rush of air, a billow of cloth flowered from my back. Immediately the chute checked my descent and I looked up. Triangles of red, blue, and yellow covered me like the wings of a beautiful angel. I was safe. Finally I could fly.

It was a magnificent journey back to earth, peaceful, floating above the hubbub of the world. I watched the fields and trees grow beneath me.

I landed a quarter of a mile from the jump school. The closer I got to earth, the faster it rose, and I plummeted straight toward a pine. With some frantic tugging of the guides, I managed not to land in its crown. Nonetheless the tree reached out its arms and caught me: the parachute swept across its lower limbs and snagged enough to soften my fall, but still I landed hard against the ground—feet, knees, hips, shoulders. I was home.

During spring semester I enrolled in field botany, where I met a mountain woman who smoked, toted a gun, and more or less did as she damn well pleased. We needed a hundred inflorescences for our herbariums by the end of the semester, and she and I often collected together. We found wild ginger and Solomon's seal and blue co-hosh. We explored in the national forest and camped in the Cohutta Wilderness, roasting tinfoil packets of vege-tables over outdoor fires and bathing, head and all, in the creeks. We hiked up her secret waterfall to pee over the edge and bodysurfed naked down mountain shoals.

We loved the same things—poetry and the woods. Aloud across a campfire we read Walt Whitman, and when we described the lives we wanted, our desires were the same: to live simply, close to nature, to grow and col-lect our own food, to use plants as medicines, to be as self-sufficient as possible. She was older by ten years, returning to school with a renewed drive to follow her dreams. She was married to a trucker named Dorsey who made fun of her "roots and sticks" and torpedoed her feminist ideas, like him helping with cooking. "If I had intended to eat sandwiches all the time," he said, "there would have been no point in getting married."

The university was made for molecular biology and philosophy, for mathematics and Latin—those disciplines *are* useful—but what I needed to learn just as much was life. The first course I scheduled was swimming, where I fi-nally learned to float, to trawl, to backstroke, to sidestroke. I took folklore and astronomy, with night classes in the ob-servatory. I took spring flora. I took writing classes and fiddle lessons. Often my friend and I prowled through the natural-foods store, learning about whole wheat, herbs, brown rice, kefir, and ginseng. I researched beekeeping, chickens and goats, organic gardening, homesteading in Alaska, the peace corps, recycling. I collected maps and bonnet patterns and directions for drying fruit.

If a subject couldn't be applied, I wasn't interested.

I slept under the stars and ventured out alone to hunt a certain place where lady's slipper and trailing arbutus was said to bloom. It was as if my spirit had suddenly been let free. Nature was the other world. It claimed me.

At the same time I completed course work required of me and did well. I studied. I wrote letters to my brothers and parents back home, and returned to Baxley, six hours south, for holidays and occasional long weekends. I became student chair of cultural affairs at the college, responsible for cultural events—movies on Friday nights in the auditorium, art shows in the gallery, visiting plays.

My coup was James Dickey's visit to campus. I wrote and told him we couldn't offer much honorarium, just $750, but I could take him out in the woods he'd written so much about, if he would do a reading. The advisers agreed, despite skepticism: who would come hear a poet?

"If you pan an additional $250 out of those skinny creeks up there, offering me $1,000 in all," he wrote back, "which is one seventh of my usual fee, I will come. I love that area and hope to be buried there. My poem 'Winter Trout' is about trout in Rock Creek."

We put notices in newspaper and on radio stations for miles around, all through those isolated mountain communities. We papered the campus and the town with fliers and even made table tents for the cafeteria, announcing that the author of *Deliverance*, the man who'd played sheriff, would be speaking.

An hour before the reading, the auditorium began to fill, and by seven it was packed. Over a thousand people had come to hear James Dickey, poet, read.

The next morning some of us picked him up at the

motel and drove him into the wilderness. We went to Rock Creek, where we hiked to a footbridge, then built a fire by water's edge and had a picnic. We drank beer and talked. Dickey carried a small pair of binoculars around his neck, to look at birds. I'd never seen that. He talked to the locals we met like a starving man, old men and women fishing the creek with corn kernels, and all day he seemed happy.

Although I'd read Thoreau and Marjorie Kinnan Rawlings in high school, that first year away I heard the word *environmentalism* and first saw it in action. I had just arrived at North Georgia. One dewy morning in early September, outside the Biology Building, a hand-lettered sign hung from a tree that was threatened with being cut.

"Woodman, spare that tree," it read. "For in my youth it shaded me. And I'll protect it now."

I knew George Pope Morris's poem, but in my history I'd not heard of a person who took it literally. The idea of caring for a tree for the tree's sake was so sentimental it was foreign, and the longer I thought about it, the more I admired whoever had lettered those words late at night in a biology lab and tied them furtively under cover of darkness to the threatened maple. One simple act turned my thinking, made me wish I knew myself better and wasn't gripped with fear when I spoke.

What I learned at North Georgia was the direction my life would go. I also learned that I would never lose the tug of the past on my life.

A few years ago, between visits home, a man named Sellers who lives in the northern part of Appling County, near the river, called my father.

"Mr. Frank," he said, "what was your daddy's name?"

"Charlie Joe," he replied.

"I've got a piece of tree here," said the man, "with CJ Ray carved in it and the year 1928. Right above that it has my daddy's initials and the year 1918."

"I've got it out here," continued the man. "I would've left it growing, but the pulpwooders are coming, and I was afraid they'd get too close to it. If you can cut it in two without hurting my daddy's name, you can have the piece with your daddy's name."

My father took the length of hardwood to the high school and rived it with the shop teacher's band saw. The length of wood sits in his parlor now, with the fancy chairs and a crash of china figurines, where I kneel to look at it.

When my father mentions how long the carving must have taken, I smile.

CJ Ray, 1928, it says. Scars bear testimony to occupation, to event, but not to circumstances; to phenomena but not conditions. They are evidence only, not details. In which direction was he walking? Were the soles of his boots intact? Was the knife fresh-sharpened? What did he witness? I want to know more than the fact that my grandfather was in the Altamaha floodplain in 1928.

"Your Mama and I decided to do the same thing," Daddy told me. I raised my eyebrows. "Carve our initials," he said. After all four children were grown, Mama and Daddy had bought a piece of timberland in Telfair County, where Grandpa was born. They began to spend time there, driving over on Sunday afternoons, planting trees, walking the acreage looking for deer, noticing the gopher burrows. Without children to support, some part of my father had turned back to land. It took Daddy an hour to carve this in a tree: *F.D.R. & L.A.R., 1986.*

"What kind of tree was it?" I asked him. He didn't know. A hardwood.

"Is it easy to find?" I asked, thinking I might be rescuing the memory of them one day from the woods. Scars turn into trinkets you keep, wounds the herons and pitcher plants heal.

"Oh, yes," he said. "Close to the road."

Second Coming

Through the acres of wrecks she came
With a wrench in her hand,

Through dust where the blacksnake dies
Of boredom, and the beetle knows
The compost has no more life.

—James Dickey, "Cherrylog Road"

When my father bought a ten-acre lowland out U.S. 1 north on the outskirts of Baxley, Georgia, intending to use it as a junkyard, it had already been logged. So the unhitching of the first junker wasn't so much a travesty as it was a monument to my deepest regret.

Birding in the junkyard now, one finds nothing very unusual: cardinals, brown thrashers, red-winged black-birds, crows. They eat the ripe elderberries and the mosquitoes that rise from the environs of foundered vehicles. Although I did not as a child know their name, Carolina wrens nest in the old cars, from which anoles and snakes come crawling. Field mice birth pink babies into shredded foam under back seats.

But where are the eastern bluebirds, winter chicka-
dees, yellow-rumped warblers, white-eyed vireos? Where
are tree swallows and savanna sparrows? Where is yellow
colic root and swamp coreopsis? Where is bird's-foot vio-
let and blue-eyed grass? Where are meadowlarks? River
swamp frogs and sweet bay magnolias should be there, an
alligator or two. What happened to the cougar and the
red wolf?

Sometimes I dream of restoring the junkyard to the
ecosystem it was when Hernando de Soto sauntered into
Georgia, looking for wealth but unable to recognize it.
Because it is a lowland, perhaps transitional to a bog, slash
pine would have dominated. Slash pines still grow here
and there, as well as other flora native to a wet pinewood:
hatpins, sundews, gallberry. I dream about it the way my
brother dreams of restoring the '58 Studebaker, a fender
at a time.

Eighty to 95 percent of the metals of vehicles of that
era are recyclable, but what do you do with the gas tanks?
What about heavy metal accumulations in the soil, lead
contamination, battery acid leaks, the veins of spilled oil
and gasoline? The topsoil would have to be scraped away:
where would it go? What about the rubber, plastic, and
broken glass? Would we haul it all to the county dump?

It might take a lifetime, one spent undoing. It might
require even my son's lifetime. And where would we find
all the replacement parts for this piece of wasted earth?
Yet, might they not come, slowly, very slowly?

A junkyard is a wilderness. Both are devotees of decay.
The nature of both is random order, the odd occurrence
and juxtaposition of miscellany, backed by a semblance

of method. Walk through a junkyard and you'll see some of the schemes a wilderness takes—Fords in one section, Dodges in another, or older models farthest from the house—so a brief logic of ecology can be found.

In the same way, an ecosystem makes sense: the canebrakes, the cypress domes. Pine trees regenerate in an indeterminate fashion, randomly here and there where seeds have fallen, but also with some predictability. Sunlight and moisture must be sufficient for germination, as where a fallen tree has made a hole in the canopy, after a rain. This, too, is order.

Without fail in a junkyard you encounter the unexpected—a doll's head, bodyless; a bike with no handlebars; a cache of wheat pennies; thirty feet of copper pipe; a boxy '58 Edsel. Likewise, in the middle of Tate's Hell Swamp you might look unexpectedly into the brown eyes of a barred owl ten feet away or come upon a purple stretch of carnivorous bladderworts in bloom, their BB-sized bladders full of aquatic microorganisms.

In junkyard as in wilderness there is danger: shards of glass, leaning jacks, weak chains; or rattlesnakes, avalanches, polar bears. In one as in the other you expect the creativity of the random, how the twisted metal protrudes like limbs, the cars dumped at acute, right, and obtuse angles, how the driveways are creeks and rivers.

This from my brother Dell:

> *There is a place in the old junkyard that, when I encounter it, turns magical. I become a future savage, half-naked, silently creeping through the dense canopy of trees and scrub. A feeling of dread increases with each step but curiosity draws me on. My footsteps falter but never completely*

stop. Suddenly I see mammoth beasts, eyes staring sightless forward. I see huge shining teeth in these monsters. As I move my hand gently among their flanks, I realize that I am in a graveyard speckled with dead prehistoric creatures. I am filled with awe. I can only speculate about their lives, imagine them roaring about and shudder at what they fed on. I know that this is hallowed ground and I remember that this place was spoken of in soft mutterings of the old ones, long dead, around the fires at night. But hunger pangs drive me on, for beyond this place are the animals that clothe and feed me. As I grope the haft of the spear and prepare to leave, I wonder if the pangs are from hunger or from a sense of loss.

Pine lilies don't grow in the junkyard anymore, nor showy orchis, and I've never seen a Bachman's sparrow flitting amid the junk. I'd like to. I have a dream for my homeland. I dream we can bring back the longleaf pine forests, along with the sandhills and the savannas, starting now and that we can bring back all the herbs and trees and wild animals, the ones not irretrievably lost, which deserve an existence apart from slavery to our own.

Afterword: Promised Land

When we consider what is happening to our forests—and to the birds, reptiles, and insects that live there—we must think also of ourselves. Culture springs from the actions of people in a landscape, and what we, especially Southerners, are watching is a daily erosion of unique folkways as our native ecosystems and all their inhabitants disappear. Our culture is tied to the longleaf pine forest that produced us, that has sheltered us, that we occupy. The forest keeps disappearing, disappearing, sold off, stolen.

In the midst of new uncertainties in the world, including global economics and a frenzy of technology, we look around and see that the landscape that defined us no longer exists or that its form is altered so dramatically we don't recognize it as our own. Animals that adapted as we adapted to place dwindle and die out. The rivers that have been lifelines are polluted by radioactive waste. Where do we turn? To what then do we look for meaning and consolation and hope?

We recognize that the loss of our forests—which is to say of health, of culture, of heritage, of beauty, of the infinite hopefulness of a virgin forest where time stalls—is a

271

loss we all share. All of our names are written on the deed to rapacity. When we log and destroy and cut and pave and replace and kill, we steal from each other and from ourselves. We swipe from our past and degrade our future.

We don't mind growing trees in the South; it's a good place for silviculture, sunny and watery, with a growing season to make a Yankee gardener weep. What we mind is that all of our trees are being taken. We want more than 1 percent natural stands of longleaf. We know a pine plantation is not a forest, and the wholesale conversion to monocultures is unacceptable to us.

We Southerners are a people fighting again for our country, defending the last remaining stands of real forest. Although we love to frolic, the time has come to fight. We must fight.

In new rebellion we stand together, black and white, urbanite and farmer, workers all, in keeping Dixie. We are a patient people who for generations have not been ousted from this land, and we are willing to fight for the birthright of our children's children and their children's children, to be of a place, in all ways, for all time. What is left is not enough. When we say the South will rise again we can mean that we will allow the cutover forests to return to their former grandeur and pine plantations to grow wild.

The whippoorwill is calling from the edge.

There Is a Miracle for You If You Keep Holding On

I will rise from my grave with the hunger of wildcat,
wings of kestrel, and with possession of my granddaughter's granddaughter, to see what we have lost
returned. My heart will be a cistern brimming with
rainwater—drinkable rain. She will not know my
name, though she bears the new forest about her, the
forest so grand. She will have heard whooping cranes
witnessing endless sky. While around her the forest
I longed all my short life to see winks and slips and
shimmers and thumps. Mutes and musks and lights.
She will walk through it with the azure-bodied
eagerness of damselfly. *My child,* I will try to call to her.
*My child. I have risen from the old cemetery buried in the forest
where your people are laid. Where once a golf course began.
That was houses and fields long, long ago.* She will be yet
a child, and may not hear me. Perhaps I will not speak
at all but follow her through a heraldry of longleaf,
seeking for the course of a day the peace of pine
warblers. And in the evening of that blessed day,
I will lay to rest this implacable longing.

Appendixes

Recently Extinct Species

Endangered Species

Species Proposed for Endangered Status

Longleaf Resources

RECENTLY EXTINCT SPECIES

Animal and plant species associated
with uplands in the southern coastal plain.

Carolina parakeet
Ivory-billed woodpecker
Passenger pigeon
Woodland bison
Red wolf
King vulture

Endangered Species

Animal and plant species associated with longleaf pine or
wiregrass communities in the southern coastal plain.

Plants

Apalachicola rosemary
Pigeon-wing
Beautiful pawpaw
Rugel's pawpaw
Scrub mint
Scrub buckwheat
Harper's beauty
Rough-leaf loosestrife
Britton's beargrass
Godfrey's butterwort
Chapman's rhododendron
Michaux's sumac
Green pitcherplant
Chaffseed
Gentian pinkroot
Cooley's meadowrue
Clasping warea
Carter's warea

Reptiles

Gopher tortoise
Sand skink
Indigo snake
Blue-tailed mole skink

BIRDS
Mississippi sandhill crane
Bald eagle
Florida scrub jay
Red-cockaded woodpecker

MAMMALS
Florida panther

Species Proposed for Endangered Status

Animal and plant species associated with longleaf pine or wiregrass communities in the southern coastal plain.

Plants

Incised groovebur
Carolina lead-plant
Georgia lead-plant
Southern three-awned grass
Southern milkweed
Chapman's aster
Coyote-thistle aster
Pinewoods aster
Sandhills milkvetch
Purple balduina
Hairy wild-indigo
Scare-weed
Ashe's savory
Sand-grass
Piedmont jointgrass
Large-flowered rosemary
Tropical waxweed
Umbrella sedge
Dwarf burhead
Telephus spurge
Wiregrass gentian
Florida beardgrass
Hartwrightia
Mock pennyroyal
Spider lily
Thick-leaved water willow
White-wicky
Tiny bog buttons
Pine pinweed
Godfrey's blazing star
Slender gay-feather
Panhandle lily

Bog spicebush
Large-fruited flax
Harper's grooved-yellow flax
West's flax
Boykin's lobelia
White birds-in-a-nest
Carolina bogmint
Southern marshallia
Bog asphodel
Fall-flowering ixia
Florida beargrass
Savanna cowbane
Naked-stemmed panic grass
Carolina grass-of-parnassus
Wavyleaf wild quinine
Chapman's butterwort
Bent golden-aster
Pineland plantain
Wild coco, eulophia
Sandhills pixie-moss
St. John's Susan, yellow coneflower
Bog coneflower
White-top pitcherplant
Wherry's pitcherplant
Florida skullcap
Scarlet catchfly
Carolina goldenrod
Spring-flowering goldenrod
Wireleaf dropseed
Pickering's morning-glory
Pineland hoary pea
Smooth bog-asphodel
Shinner's false-foxglove
Least trillium
Chapman's crownbeard
Variable-leaf crownbeard
Drummond's yellow-eyed grass
Harper's yellow-eyed grass

INSECTS

Buchholz's dart moth
Aphodius tortoise commensal scarab beetle
Arogos skipper
Copris tortoise commensal scarab beetle
Sandhills clubtail dragonfly
Spiny Florida sandhill scarab beetle
Prairie mole cricket
Mitchell's satyr
Onthophagus tortoise commensal scarab beetle
Carter's noctuid moth

AMPHIBIANS

Flatwoods salamander
Gopher frog
Carolina gopher frog
Dusky gopher frog

REPTILES

Gopher tortoise
Florida scrub lizard
Southern hognose snake
Black pine snake
Northern pine snake
Florida pine snake
Short-tailed snake

BIRDS

Southeastern American kestrel
Loggerhead shrike
Bachman's sparrow
Henslow's sparrow

MAMMALS

Florida weasel
Florida black bear
Florida mouse
Sherman's fox squirrel

Longleaf Resources

Organizations working to assist longleaf pine forests
and resources for further consultation.

Organizations

Coastal Plains Institute

• A nonprofit environmental research and education
organization since 1984.
1313 North Duval Street
Tallahassee, FL 32303-5512
(850) 681-6208 / phone
(850) 681-6123 / fax
means@bio.fsu.edu

The Dogwood Alliance

• Network of grassroots organizations in the Southeast who
are working together to stop industrial clearcutting for chip
mills.
P. O. Box 1598
Brevard, NC 28712
(828) 883-5889 / phone
(828) 883-5826 / fax
info@dogwoodalliance.org
www.dogwoodalliance.org

Gopher Tortoise Council

c/o Florida Museum of Natural History
University of Florida
P. O. Box 117800
Gainesville, FL 32611-7800
www.flmnh.ufl.edu/natsci/herpetology/gtc.htm

Longleaf Alliance

• Alliance to promote the ecological and economic values of
longleaf pine ecosystems by encouraging better management
practices in order to reverse the decline.

Solon Dixon Forestry Education Center
Route 7, Box 131
Andalusia, AL 36420
(334) 222-7779 / phone
(334) 222-0581 / fax

The Nature Conservancy of Georgia Longleaf Initiative
P. O. Box 484
Darien, GA 31305
(912) 437-2161 / phone
(912) 437-5368 / fax

North Carolina Nature Conservancy
4011 University Drive, Suite 201
Durham, NC 27636
(919) 403-8558 / phone

School of Forestry
Auburn University, AL 36849

South Carolina Nature Conservancy
P. O. Box 5475
Columbia, SC 29250
(803) 254-9049 / phone

Tall Timbers Research Station
Route 1, Box 678
Tallahassee, FL 32312
(850) 893-4153 / phone

BOOKS AND VIDEOS
Discovering Alabama Video #29
• Public television production emphasizing the history of long-leaf and current initiatives for recovery. Includes a teacher's guide.

Alabama Museum of Natural History
Box 870340
Tuscaloosa, AL 35487
(205) 348-2039 / phone

Discovering Alabama Video #30
- Public television production highlighting research and management efforts on a variety of species associated with longleaf. Includes a teacher's guide.

Alabama Museum of Natural History
Box 870340
Tuscaloosa, AL 35487
(205) 348-2039 / phone

"Remnants of a Forest"
- A twenty-seven minute video about the longleaf pine in Georgia, available for purchase or on loan.

Georgia Department of Natural Resources
205 Butler Street Southeast, Suite 1354
Atlanta, GA 30334
(404) 657-9851 / phone

"Stewardship of Longleaf Pine Forests: A Guide for Landowners"
- Booklet published by the Longleaf Alliance.

Longleaf Alliance
Solon Dixon Forestry Education Center
Route 7, Box 131
Andalusia, AL 36420
(334) 222-7779 / phone
(334) 222-0581 / fax

"A Working Forest: A Landowner's Guide for Growing Longleaf Pine in the Carolina Sandhills"
- Booklet published by Sandhills Area Land Trust.

Sandhills Area Land Trust
P. O. Box 1032
Southern Pines, NC 28388
(910) 695-4323 / phone

ACKNOWLEDGMENTS

For unflagging encouragement of the highest order in the making of this book, start to finish, I thank my dear friend Susan Cerulean, whose love and good thinking sustain me. Rick Bass has been a wellspring of inspiration; his work as a prophet of wildness has been a light I have walked toward, his friendship something holy. I thank my editor, Emilie Buchwald, for her brilliance and generosity and for leading a life exemplary in its service to humanity.

Characters in this book are real people, and with love I acknowledge their work in the world and ask that their privacy be respected: my good parents, Franklin and Lee Ada Ray; my brothers, Dell and Stephen Ray; and my sister, Kay Amsler. A million thanks to my beloved Milton N. Hopkins.

Bill Kittredge, the first to see the manuscript, steered me away from many a cliff. Wendell Berry, Hank Harrington, and Peter Matthiessen also kindly read and edited parts and I thank them. The help of Patricia Traxler and Pattiann Rogers has been essential.

Thanks also to Bill Belleville, Andrea Blount, Percy Branch, Todd Engstrom, Dan Flores, Manley Fuller, Gary Graham, Angus and Eloise Gholson, Lizzie Grossman, my agent Jennifer Hengen, Joe Kipphut, Stephen J. Lyons, Leon and Julie Neel, Barbara Ras, Jean Tyre Ray, Alan J. Roach, Annick Smith, Frankie Snow, Melissa Walker, and Mick Womersley; as well as the staff of Appling County Public Library and the University of Montana's Mansfield Library. The University of Montana provided scholarships and assistantships that allowed time to write. Thank you.

JANISSE RAY was born in 1962 and is a native of the coastal plains of southern Georgia. *Naming the Unseen,* her chapbook of poetry about biology and place, won the 1996 Merriam-Frontier Award from the University of Montana, where Ray earned an MFA in creative writing in 1997. A naturalist and environmental activist, Janisse has published essays and poems in such newspapers and magazines as *Wild Earth, Hope, Tallahassee Democrat, Missoula Independent, Orion, Florida Wildlife,* and *Georgia Wildlife,* among others. She lives on a family farm in Baxley with her son.

More Books on The World As Home
from Milkweed Editions

To order books or for more information,
contact Milkweed at (800) 520-6455
or visit our website (www.milkweed.org).

Brown Dog of the Yaak:
Essays on Art and Activism
RICK BASS

Boundary Waters:
The Grace of the Wild
PAUL GRUCHOW

Grass Roots:
The Universe of Home
PAUL GRUCHOW

The Necessity of Empty Places
PAUL GRUCHOW

A Sense of the Morning:
Field Notes of a Born Observer
DAVID BRENDAN HOPES

Taking Care:
Thoughts on Storytelling and Belief
WILLIAM KITTREDGE

The Dream of the Marsh Wren:
Writing As Reciprocal Creation
PATTIANN ROGERS

The Country of Language
SCOTT RUSSELL SANDERS

The Book of the Tongass
EDITED BY CAROLYN SERVID AND DONALD SNOW

Homestead
ANNICK SMITH

Testimony:
Writers of the West Speak On Behalf of Utah Wilderness
COMPILED BY STEPHEN TRIMBLE
AND TERRY TEMPEST WILLIAMS

Other books of interest to
The World As Home reader:

ESSAYS

The Heart Can Be Filled Anywhere on Earth:
Minneota, Minnesota
BILL HOLM

Shedding Life:
Disease, Politics, and Other Human Conditions
MIROSLAV HOLUB

CHILDREN'S NOVELS

No Place
KAY HAUGAARD

The Monkey Thief
AILEEN KILGORE HENDERSON

Treasure of Panther Peak
AILEEN KILGORE HENDERSON

The Dog with Golden Eyes
FRANCES WILBUR

ANTHOLOGIES

Sacred Ground:
Writings about Home
EDITED BY BARBARA BONNER

Verse and Universe:
Poems about Science and Mathematics
EDITED BY KURT BROWN

POETRY

Boxelder Bug Variations
BILL HOLM

Butterfly Effect
HARRY HUMES

Eating Bread and Honey
PATTIANN ROGERS

Firekeeper:
New and Selected Poems
PATTIANN ROGERS

The World As Home, the nonfiction publishing program of Milkweed Editions, is dedicated to exploring our relationship to the natural world. Not espousing any particular environmentalist or political agenda, these books are a forum for distinctive literary writing that not only alerts the reader to vital issues but offers personal testimonies to living harmoniously with other species in urban, rural, and wilderness communities.

Milkweed Editions publishes with the intention of making a humane impact on society, in the belief that literature is a transformative art uniquely able to convey the essential experiences of the human heart and spirit. To that end, Milkweed publishes distinctive voices of literary merit in handsomely designed, visually dynamic books, exploring the ethical, cultural, and esthetic issues that free societies need continually to address. Milkweed Editions is a not-for-profit press.